TECHCAREERS:

Aviation Pilots

Jodi Wheatley

TSTC Publishing

ISBN 978-1-934302-89-7 (softback)
ISBN 978-1-936603-11-4 (ebook)

TSTC Publishing
Texas State Technical College Waco
3801 Campus Drive
Waco, Texas 76705

http://publishing.tstc.edu/

Publisher: Mark Long
Art director: Stacie Buterbaugh
Editor: Ana Wraight
Marketing: Sheila Boggess
Sales: Wes Lowe
Office coordinator: Melanie Peterson
Graphics Specialist: Grace Arsiaga
Indexing: Michelle Graye (indexing@yahoo.com)
Printing production: Data Reproductions

Manufactured in the United States of America

First edition

Publisher's Cataloging-in-Publication
(Provided by Quality Books, Inc.)

Wheatley, Jodi.
Aviation pilots / Jodi Wheatley. -- 1st ed.
p. cm. -- (TechCareers)
Includes index.
ISBN-13: 978-1-934302-89-7 (softback)
ISBN-13: 978-1-936603-11-4 (ebook)

1. Air pilots. 2. Aeronautics--Vocational guidance.
I. Title. II. Series: TechCareers.

TL561.W44 2012 629.13'023
QBI11-600197

Table of Contents

Commonly Used Abbreviations

AAS	Associate of Applied Science
AOPA	Airplane Owners and Pilots Association
APT	Aircraft Pilot Training
ASA	American Southeast Airlines
ATP	Airline Transport Pilot
BU	Baylor University
CFI	Certified Flight Instructor
CFII	Certified Flight Instructor Instrument
IFR	Instrument Flight Rules
FAA	Federal Aviation Administration
FAR	Federal Aviation Regulations
FO	First Officer (also called copilot)
IFR	Instrument Flight Rules
MEI	Multi-Engine Instructor
NIFA	National Intercollegiate Flying Association
OES	Occupational Employment Statistics
OOH	Occupational Outlook Handbook
OTS	Officer Training School
P61, 141, 142	The parts of FAR covering pilot certification and flight school operations: the pilot certification and standard flight school (Part 61), the integrated curriculum-type school (Part 141) requiring slightly fewer flying hours, and a new Part 142 program allowing replacement of more flight time with advanced flight simulators
REC	Recreational
SAFECON	Safety and Flight Evaluation Conference
SIC	Second-in-Command
TAMU-CT	Texas A&M University – Central Texas
TFB	Texas Farm Bureau

TSC	Texas State Credit
TSTC	Texas State Technical College
WAI	Women in Aviation, International

CHAPTER ONE

Aviation Pilot Careers

"When once you have tasted flight, you will forever walk the earth with your eyes turned skyward..." – Leonardo da Vinci

For thousands of years, humans have watched the birds and yearned, even ached, to fly. In the late eighteenth century, lighter-than-air human flight was achieved with primitive hot air balloons. On December 17, 1903, the Wright brothers famously made the first truly successful heavier-than-air powered aircraft flight at a speed of 6.8 mph.

Although it took a long time to achieve the beginning of human flight, from that point on, aviation made giant leaps forward approximately every decade.

- Less than ten years after the Wright brothers' flight, the first aircraft manufacturer opened for business, and the first jet-propelled aircraft was tested.
- Less than ten years later, airplanes played a significant role in World War I.
- Less than ten years later, the first airline was founded, and a flight speed record of 407.5 mph was set.
- Less than ten years later, Charles Lindergh completed the first solo, nonstop crossing of the Atlantic Ocean.

- A little more than ten years later, United States and British bombers, fighters, and supply planes launched into the air all over the world, making aircraft an indispensable part of the battles of World War II.
- Less than ten years later, the sound barrier was broken by General Chuck Yeager.
- Less than ten years later, the Havilland DH 106 Comet became the first commercial jet airliner.
- Less than ten years later, the first manned aeronautical mission launched into space, and less than ten years after that, man flew to the moon and back.
- Less than ten years later, the supersonic passenger jet Concorde flew high enough to see the curvature of the Earth, and less than ten years after that, two pilots flew around the world nonstop and without refueling.
- Less than ten years later, stealth bombers "owned the nights" during the first Gulf War.
- Just over ten years after that, *SpaceShipOne* completed the first privately funded spaceflight, only 101 years after the Wright brothers' first heavier-than-air powered flight.

Not only does aviation continue to make major advances each decade, but according to the *Joseph T. Nall Report* published annually by the Aircraft Owners and Pilots Association (AOPA), the number of aircraft in use roughly doubles every ten years. In spite of short-term ups and downs, this increase means the need for new commercial pilots continues to grow.

Humans still want to fly, and happily, the opportunities for men and women to do so as a career are consistently increasing.

Industry Organization

Airlines transport passengers and freight either over regularly scheduled routes or on charter routes individually designed for a specific group of travelers or cargo. Several classes of airlines operate in the United States. Mainline air carriers use large passenger jets with more than ninety seats. Regional carriers use smaller piston, turboprop, and regional aircraft with up to ninety seats. There also are all-cargo carriers, which use aircraft designed exclusively to carry freight.

Some mainline carriers are called "network carriers." These carriers use a centrally located airport as a hub, where they receive a large number of their flights from many locations and where passengers can transfer from one flight to another. Other mainline carriers operate as low-cost carriers. Offering only flights between limited numbers of cities, low-cost carriers do not typically use hubs. In recent years, these companies have expanded their customer base to include more business travelers. They are the fastest-growing segment of commercial aviation.

Profile: *Shelley Meline, Professional Pilot*

Now a veteran airline pilot with AirTran Airways, Shelley Meline remembers the exact day she decided to become a pilot. At the age of nine, she accompanied her father to work at the Texas Air Expo. He was in charge of a group of golf carts used for running errands and making deliveries around the host airfield. She rode with him as they delivered water to pilots who were demonstrating their aircraft. That evening, she told her parents she wanted to be a pilot, and she never let go of the dream.

In high school, Meline began taking lessons at a local flight school, earning her private pilot's certificate in 1996. For further training, she researched colleges with aviation programs, eventually settling on Baylor University's (BU) four-year Aviation Science degree.

Although she was "always a real BU fan," she says, choosing it for her education was not a given. Her research convinced her BU had just as much to offer as the other schools she had researched, and some of the others had higher tuition. It was also close to home.

With student loans in hand, she tackled the challenge of BU's program. "It was really good," she says. "It was pretty well balanced between the academics and the time for the flying courses, which were conducted at Texas State Technical College (TSTC) Waco, since BU has no aircraft of its own for flight training." She describes the program at the time as heavy in math and computer science courses she had to take, which now "I'm glad I did," she admits. The hardest challenge she encountered at college was paying for the flying. She got a job to help cover the expenses, took flight classes during the summer, and graduated in four and a half years.

Her first job out of college was as a flight instructor at a nearby P64 flight school, followed by almost a year as a flight instructor at TSTC Waco. "I really liked the P141 school for flight instructing. I got more students, and they were at different levels. I enjoyed getting to know them and taking some from the beginning up through the more advanced certifications." Helping others succeed in the field she loves is very satisfying for her. "I really enjoyed the teaching, although it demanded a lot of hours. I keep up with many of my former students. It's neat to see how they're doing." Some of her former students now work with her at AirTran Airways.

Meline's next job after instructing was with Mesaba, the longest-flying regional airline in the United States, a subsidiary of Delta and a connection carrier for both Delta and Northwest Airlines. She flew for Mesaba as a first officer (FO) on a thirty-four-passenger SAAB SF340. Based in Detroit for more than two years, "I got some good winter flying experience," she says, laughing. She also met her future husband, Michael, there.

She now works at AirTran Airways, a low-cost carrier similar to Southwest and ranked as a major airline on the basis of its revenue. Meline usually flies a 737 now, which makes trips from coast-to-coast and to destinations in the Caribbean and Mexico. After Meline joined AirTran, Michael also came aboard, making her senior to him at their new employer, although he had been senior to her at Mesaba. This kind of position reversal is common among pilots, according to Meline. Pilots who are older than she and were captains she flew with as FO at Mesaba lost their jobs in the slow economy and are now at AirTran as FOs under her as captain.

Meline currently works a schedule that gives her about fifteen days off a month. She and Michael are frequently able to work similar schedules, doing mostly four-day trips followed by three or four days at home. Pilots at AirTran "bid" for schedules, according to Meline. Every month the next month's schedules are posted, and pilots rank the flights they prefer from most- to least-desired. Schedules are awarded based on the seniority of the pilots in the company and in flight hours on the different types of planes used by the airline. This gives the pilot quite a bit of flexibility and control over their flight schedules. Some pilots, according to Meline, prefer to remain senior FOs and have more choice over their schedules, rather than becoming junior captains and having to be on reserve duty until they work their way up in seniority again. Reserve is the first level for the newest, most junior captains and FOs, she says, and requires them to be available for flights for twelve- to fourteen-hour stretches. If they get a callout, they must be at the airport ready to fly in two hours. It is very unpredictable. One month they might get lots of hours and the next month none, but it is the first step on the ladder at most airlines.

Throughout her schooling and career so far, Meline says she has never experienced any discrimination or pressure because she is a woman. "I felt like I was regarded pretty much the same as other pilots. They treated me just like 'one of the guys'." Hiring and training in new aircraft were also equal-access for her. She says that the only place she has really noticed any real gender bias is from the flying public – some people seem surprised to see a young woman serving as captain.

All in all, Meline feels very lucky in her career. "All my employers have been great to work for," she says. "I've hit the right places at the right times. My employers have provided the training for me for the planes they had me fly. I've never encountered any furloughs.

You've got to stick with it. A lot of it is timing, being there when the jobs open up. I remember as a student wondering if it would ever happen for me."

Someone gave her some advice a long time ago: "Don't go into it for the money. Do it because you love flying." She considers that to be good advice still, as the airline industry is very cyclical. New pilots must be prepared for possible furloughs when the economy bottoms out. Even when the economy is good, a lot of time is spent in hotel rooms, away from home. "The commuter airlines are very challenging financially," she warns. She was able to pay off her student loans while working at one, but says it was "really tight." Now, however, she is seeing some people stick with the commuter airlines longer, staying on and building enough seniority to make a good living there. In fact, she is seeing a lot less movement in general in the industry.

Her advice to students is: "Dedication, dedication, dedication! Be persistent, and make sure it's something you love. If it's something you love, you'll have a great career no matter what happens. I fly with people every day who had other careers and became pilots because they always wanted to fly." Overall, she feels that, just like her, the pilots she knows are one of the happiest groups of workers she has ever encountered.

Regional carriers emphasize connecting smaller communities with larger cities and hubs. They operate short- and medium-haul scheduled airline services. Some are subsidiaries of larger carriers, but many independently owned regional carriers contract their services to the mainline carriers.

Cargo carriers do not provide "door-to-door service," but they do provide air transport from an airport near the cargo's origin to an airport near the cargo's destination. Companies that provide door-to-door delivery of cargo, whether it is across town or overseas, are classified by the U.S. Department of Labor as couriers and messengers, even though they may have planes and hire pilots.

Employment Outlook

When the economy is doing well, many people fly for business and leisure, and many businesses use overnight delivery services for packages. Even when the economy is slow, however, planes still fly for both business and recreational purposes, though in fewer numbers at times. Whatever the economic status of the country, all of these activities require pilots to fly planes.

According to the *Occupational Outlook Handbook* (OOH), published by the U.S. Department of Labor: "Civilian aircraft pilots and flight engineers held about 116,000 jobs in 2008. About 76,800 worked as airline pilots, copilots, and flight engineers. The rest were commercial pilots who worked as flight instructors at local airports or for large businesses that fly company cargo and executives in their own airplanes or helicopters. Some commercial pilots flew small planes for air-taxi companies, usually to or from lightly traveled airports not served by major airlines. Others worked for a variety of businesses, performing tasks such as dusting crops, inspecting pipelines, or conducting sightseeing trips. Pilots are located across the country, but airline pilots usually are based near major metropolitan airports or airports operating as hubs for the major airlines."

Profile: *Mike Smith, Professional Pilot*

In grade school, Mike Smith's classroom dreams of flying were often interrupted by the sudden smack of the teacher's ruler on his hand. Scolding him for his distraction, she asked him, "Where are you going to find a job where you can look out the window all day?" It took years of dedication and focus, but Smith eventually made those dreams come true. Now as a flight captain for Southwest Airlines, he says he gets to do just what his teacher said: he looks out the window all day from the cockpit of the Boeing 737.

Inspired by the exciting air show aerobatics of the Thunderbirds and Blue Angels he watched from the hood of the family car, he was determined to fly for the military. Smith admits he did poorly in high school, mostly because he did not study. He played football well, though, and was recruited to play for the New Mexico Military Institute. As a result, he experienced his first real flight in a 727. "It was just like in my dreams," he says. At that point he realized he would never be able to fly if he didn't graduate from college. Buckling down to the task, he started making As and Bs.

After two years, he realized that, although he loved football, playing carried a real risk of injury, which could destroy his dream to become a pilot. He decided to leave the program and pursue aviation instead. No football meant no scholarship, though. Therefore, he switched to BU and obtained an Air Force scholarship. He was commissioned with a delayed entry, on his way to his dream.

While he attended BU, however, the Vietnam War ended. By the time he finished school, the Air Force said they didn't have space for him right away. In the meantime, he worked and waited for the call that would tell him space had opened up. When the call finally came, though, it was to tell him the Air Force was no longer offering flight school.

Still determined to achieve his dream, he called the Navy and the Army, neither of which had any flying slots. Finally, a Marine Corps gunnery sergeant told him the Marines had two slots available but warned, "You can't get the paperwork done in time." It was Tuesday morning, and a certified birth certificate and notarized transcripts had to be turned in by Friday afternoon. Transcripts at the time usually took weeks to be delivered, but Smith's promise was, "I will see you Friday, Gunny!" On Friday afternoon, he walked into the Marine recruiting office and handed his completed paperwork to a surprised Gunny.

He did well at Officers Training School (OTS), and when they called out the names of the two who had earned the flight school slots, one of them was his. After an initial reaction of "There's another guy here named Mike Smith?", he realized the door to his dream had finally opened.

After ten and a half years of flying A-6 Intruders as a Marine and a couple of overseas tours, Smith decided his family needed him at home, and he resigned from the military. The airlines were not hiring, so he took a job selling industrial chemicals. "I did very well at it," he says, "and I hated every day of it." Eventually, a pilot friend called him and told him of some airlines that were hiring. Smith applied, and with the blessing of his family, went to fly for American Southeast Airlines (ASA). "I took an 80 percent pay cut to do the job I loved," he says.

During the first three years, he worked three other jobs to replace the salary he had lost: He joined the Marine reserves, steam-cleaned parking lots for a bank, and sold term life insurance, teaching customers how to invest at the same time. It was worth it to fly again, he says. On September 19, 1994, he was hired at Southwest Airlines. "That was one of the best days of my life," he states unreservedly.

Senior enough now to be able to bid for a convenient flying schedule with reasonably good results, he commutes from Central Texas to Dallas Love Field for his flights. It took perseverance and determination – tools all new pilots need to succeed – to get to this point. Without them, his own career would have derailed many times along the way. His only regret is never being able to go back to his grade school teacher and tell her that he had indeed found "a job where he gets to look out the window all day," and he considers it worth everything it took to get there.

Employment of aircraft pilots is expected to grow nearly 12 percent from 2008 to 2018, about as fast as the average for other occupations. This translates into approximately an additional 12-14,000 positions. Job growth is driven both by population growth and an expanding economy, which are expected to boost the demand for air travel in the long run. As airlines expand their capacity, the subsequent increase in the number of planes in operation and the number of flights offered should create new jobs. Regional airlines and low-cost carriers are expected to grow the fastest and present the

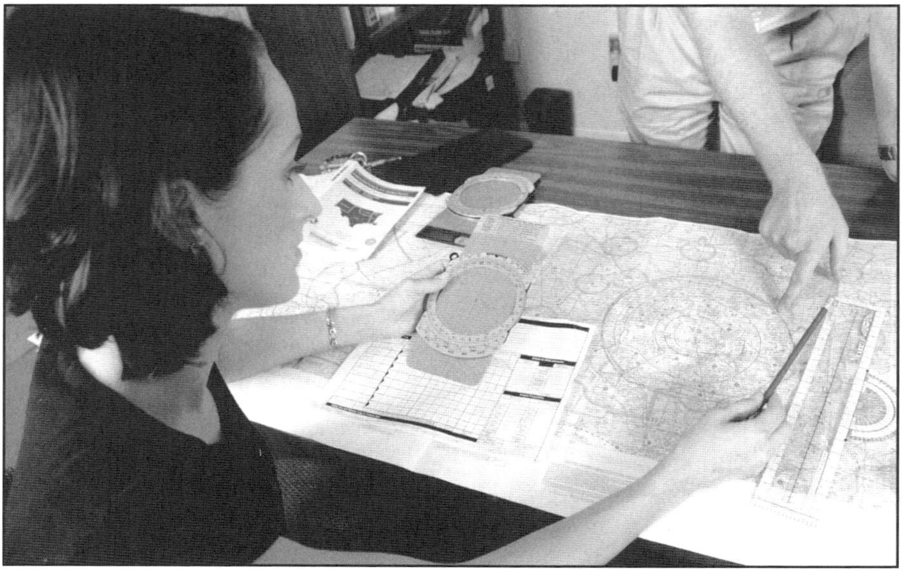

best opportunities. Although jobs are expected to increase at the major airlines also, pilots attempting to get those jobs will face strong competition. Increased call for global freight should also increase job opportunities with air cargo carriers. Pilots who have logged the greatest number of flying hours using the most sophisticated equipment usually have the best prospects.

However, as the OOH says, "Employment of pilots is sensitive to cyclical swings in the economy. During recessions, when the decline in the demand for air travel forces airlines to ground planes and curtail the number of flights, airlines may temporarily furlough some pilots." Furloughed pilots are usually first in line when jobs open up again as the economy improves.

Salary Ranges and Work Schedules

According to Ed Viera, Chief Flight Instructor for TSTC Waco, there are a few common myths about working as a professional pilot:

Myth #1: A new pilot starts out making a lot of money. "New pilots with a regional airline will start out at about $20-24,000 a year," Viera explains. Year two can increase to the $30,000 range, and year five can move into the $50,000s. Major airlines often start new pilots at about $45,000. "The days of one-airline careers and $250,000 salaries are over. Deregulation put an end to those kinds of salaries," according to Viera. Now, a pilot's career will likely include time with at least two or three different companies. There are now many smaller regional carriers flying shorter routes with smaller planes, which creates more jobs but with lower salaries due to lower consumer prices.

Myth #2: Commercial aircraft pilots live glamorous lives. In reality, Viera says, flying is a day-in, day-out job. Pilots may fly to exotic destinations, but they rarely stay there more than overnight before taking off again the next morning. Pilots may also spend quite a bit of time on reserve: waiting for the phone to ring and summon them to a relief or substitute assignment. "You have to enjoy the work itself," says Viera. "You spend a lot of time in hotel rooms and away from your family." Like most fields, you have to love flying to make it work. If you do, there is likely to be a place for you in the field.

Wages for aircraft pilots vary greatly. Factors such as working for an airline or as a corporate pilot, the type of plane, and seniority affect pay rates. New pilots are generally paid rather low wages until they log enough hours and experience to advance to new positions in larger, more complicated aircraft. According to the Payscale Research Center (payscale.com), experienced pilots with major airlines earn between $75 and $113 an hour. However, as Al Lee, Director of Quantitative Analysis for Payscale, Inc., warns in his "Dr. Salary" blog, "Be careful about turning this into an annual salary. Pilots commonly fly (and are paid for) 1000 hours or less a year, instead of the normal 2,080-hour work year (40 hours a week for 52 weeks). You could make six figures, but keep in mind, you may have to navigate through some pay cuts and labor strikes; better buckle up for a bumpy ride."

Airlines operate flights at all hours of the day and night, so many employees have irregular hours or work variable schedules, including nights, weekends, and holidays. When flight personnel are away from their home bases, airlines provide them with hotel accommodations, transportation between the airport and hotel, and an allowance for meals and expenses (called a "per diem"). Passenger airlines typically have a "bid" system for scheduling flight crews. Available routes/schedules are posted each month for the coming month, and pilots turn in their preferences as to which schedules they prefer, ranking each item on the list from first to last choice.

Schedules can range from trips of three or four days (possibly more at some airlines) to "turns," which are single-day, fly-out-and-return trips. There are also "reserve" schedules, where a pilot is on standby status for a certain number of hours and will be called in case a scheduled pilot cannot make a flight. Schedules are set by seniority, with the most senior pilots getting first choice and the most junior pilots the last. The most junior pilots usually start on reserve schedules and generally must live close to the hub to which they are assigned, as they typically receive only two hours' notice of a flight.

Profile: *Rosalyn Montgomery, Aircraft Pilot Training Student*

Rosalyn Montgomery, a twenty-three-year-old student pilot at TSTC Waco, has been flying with her father "since I was little." Both of her parents are Air Force veterans, and her father is now a commercial pilot. "I've always been connected to flying," Montgomery says. "I have a great passion for it."

Coming to TSTC Waco was a natural step for her. Both of her parents attended there: her father in Aircraft Pilot Training (APT) and her mother in Occupational Safety and Health Administration. She wanted to concentrate on her flight training while she was there. "I did two semesters at McLennan Community College for some of my basic courses, and then I started pilot training in 2006."

Besides working part-time at the Waco Family YMCA, "I'm also getting my bachelor's degree online from the University of North Texas." She was happy to find that "all of my credits from TSTC transferred for my bachelor's." She's determined to finish the degree, although she says, "I'm a real hands-on person. Online classes are hard for me. I like to be in the classroom and see what's going on."

Enjoying the classroom made TSTC Waco a good fit for Montgomery and her career plans. Taking the general college courses early freed up some of her time each semester, allowing her to concentrate on the aviation courses. She finished all of her classroom coursework before the end of 2009. Although all she needed to graduate was her flight hours, "It's hard to put a specific date on anything. Bad weather conditions can slow things down, or a plane can go out of service or be gone. A flexible attitude is important – you have to focus on your goals."

Montgomery's goal is to work her way up to a job with Southwest Airlines. "I'll work as a flight instructor, then for a regional carrier, and then try for Southwest." She chose Southwest because of its reputation as "One of the best. They respect their people, and it's family-based." She hopes to get a scholarship to accomplish the 737 rating Southwest requires, looking towards the flight schools in the Dallas-Fort Worth area that have 737 simulators to train on. She feels she will be well prepared for that training. "The work experience you get from TSTC as a student is amazing!"

Although Montgomery is confident about her ability to succeed, she is realistic about the work required for a student to achieve his or her goals in aviation. "Skill takes practice – especially this skill." She also believes in the motto of the AOPA: "A good pilot's always learning." It is a continual learning process, she explains. "Technology is always changing. They are adding new equipment to planes constantly."

Attrition is high in flight training classes, as she has seen. Out of her class of thirty, only about a third of the students were still enrolled at the end of training. The biggest problem she has seen with students is not studying. "You have to know your stuff," she says. She understands the challenge. There were times she felt she would never understand what she was trying to learn. The hardest parts of flight training for her were landing ("painting the picture in my mind of what it should look like") and instrument training ("you have to wear a hood while you fly").

She took her father's advice: be confident. "A positive, confident attitude is necessary. There will be some rough times. You have to keep going. If you really want to do it, you can." Plus, she says, students in a collegiate program aren't alone. "There are lots of people here who will give you advice. People who have a ton of experience." The benefit of persevering, according to her, is "Getting to do something you love. A lot of people think it's too late, but it's never too late to follow a dream."

The second obstacle she has seen knock out students is financial. "But there are student loans and a few scholarships." Her own situation is pretty average: She continues to acquire her required flying hours as she can afford them, working toward her goal of completing her associate degree and her commercial pilot's certificate.

A member of Women in Aviation, International (WAI), one of Montgomery's personal goals is to establish a Central Texas chapter of the organization. "I would love to see more women in the field. Eight girls started in my class, but only two are still training." She suspects a lot of girls look at the course requirements and think science and math are too hard. The math is simple, she says. She also points out that there are many people around who are willing to help.

Montgomery's advice to potential aircraft pilot trainees is this: "Do your research before going into this career. Maybe being a pilot isn't where you'll have a career, but there are other areas of aviation. Take the Discovery Flight (an inexpensive demonstration flight for people who have never flown in a small aircraft) to check your comfort level in a small plane." Most importantly, "Come ready to work. It's fun, but the work has to come first."

Federal Aviation Administration (FAA) regulations require pilots for passenger airlines fly no more than 30 hours a week, 100 hours a month, and 1000 hours per year. Flight hours are only counted from the time the cabin doors close. A significant amount of time beforehand is spent receiving information from flight dispatchers; checking the aircraft engines, controls, and instruments, as well as other systems; consulting weather forecasts; and planning the route, altitude, and other elements of the flight. None of this time, however, is included in the wage calculation. Cargo pilots and corporate pilots have different, somewhat less restrictive FAA limits, but they may fly more hours at a time with less time off between flights and have a more unpredictable flight schedule.

As reported by *Occupational Employment Statistics* (OES), published by the U.S. Department of Labor, the average wage for airline pilots (or captains) and copilots (or FOs) in May 2010 was $115,300 (10 percent of pilots made less than $55,330, and 10 percent made more than $150,480). Benefits like health insurance are usually available for pilots working for large companies, and most airlines offer free or reduced-cost travel for pilots and their families. The average wage for commercial pilots nationally was $73,490 (10 percent made less than $32,020, and 10 percent made more than

$129,580). These numbers vary by geographical region. For instance, wage rates in Texas tended to start somewhat higher than the national average and end very close to or above the national average. Careerinfonet.org reported the spread of airline pilot wages nationally and in Texas for 2009 as:

Location	Pay Period	Low	Median	High
United States	Yearly	$56,600	$106,200	$166,400
Texas	Yearly	$63,400	$108,600	$166,400

According to the annual salary report for 2009 published by *Professional Pilot* magazine, salaries for corporate pilots vary greatly according to the size and sophistication of the aircraft used, with heavy international jet captains earning the most, from $82,000 to $185,000 per year. Corporate turboprop chief pilots earned from $40,000 to $90,000 per year and captains from $37,000 to $82,000.

Career Paths and Employers

There are several career paths a professional pilot can choose. The four largest career fields are passenger airlines, freight/cargo companies (along with messenger or courier services like FedEx and UPS), corporate/business/charter services, and the military. Smaller but also viable career fields include bush piloting, scenic flight companies, emergency medical flight services, boat plane companies, pipeline companies (conducting aerial surveys and mapping), crop dusting, and law enforcement piloting (which may require extra training). Approximately one-third of professional pilots work in these fields, according to the OOH. Test piloting is another flying niche, but this usually requires additional training and expertise in areas like aircraft maintenance or engineering.

There may also be some opportunities with overseas airlines in countries that have no flight schools of their own. For international flights, English is the standard language. For flights inside a country's borders, however, the national language is used; therefore, the pilot must be able to function in that language as well. If you are considering flying in another country, research the country's certification requirements beforehand, as they may differ from United States standards.

Major airlines have many qualified applicants, which means they are able to hire only the ones they feel have the best qualifications. Most airline pilots have bachelor's degrees; consequently, to compete effectively for hiring, a new pilot is smart to acquire one, too. Two-year technical schools often offer associate degrees in aviation, which can be earned in less time and allow a new pilot to work in aviation as he or she earns a bachelor's degree.

Although there are some four-year colleges and universities that offer degrees in aviation sciences or related subjects, the academic focus of the degree does not really matter, according to Tim Compton, Assistant Department Director of the BU Aviation Sciences Department in Waco, Texas. Members of the flight deck (pilot, copilot, and flight engineer) are often tapped to participate in other duties, such as serving as flight safety officers. The program at BU includes coursework in management studies, which are valuable in these types of situations. Expertise in aviation engineering and maintenance also opens doors in some companies.

Profile: *Jim Gardner, Corporate Chief Pilot*

For Jim Gardner, growing up in Beeville, Texas, meant driving past the local airport – which had always fascinated him – on their way to the family farm every weekend. One day on a trip back to town, the family noticed flags and pennants decorating the airport. Gardner's father agreed to stop and see what was up, which turned out to be an open house, featuring things like "penny-a-pound" airplane rides and planes on display. Gardner took his first plane ride that day, which intensified his enthrallment. From then on, every time they passed the airport, Gardner wanted to stop.

When Gardner's father decided he had spent enough time away from his work at the farm, he started dropping his son off. Gardner describes himself as the iconic "kid at the airport," becoming friends with the fueler and helping him out around the place. At sixteen, he started driving on his own to the airport, and he was put to work part-time gassing planes, doing office work, sweeping up, and doing whatever else needed to be done. He started getting a few dollars and opportunities for free plane rides. If they had to ferry a plane or take a test ride, whatever the reason, Gardner took any offer to go up.

One day the owner offered Gardner free flight lessons in exchange for his work, and without hesitation, he accepted. "It was the first time I sat in the right seat, and I was really hooked." He continued swapping labor for lessons until college interrupted. Cost limited the amount of flying he could do during his time at Tarleton State University, but when he went home with a degree in agriculture, he started full-time at the airport and received his private pilot certification in 1969.

At that point, Gardner says he looked at his situation: fresh out of college, unmarried, with the Vietnam War going on. He realized he was "ripe for the picking" for the draft. He didn't really want to go into the military, but he says, "I didn't want to get drafted into the Army a lot more than I didn't want to be in the Air Force." Therefore, he went to the local Air Force recruiter, who worked out a delayed enlistment with all the paperwork done, including the aptitude test and application to OTS, except for the final signature. Ninety days later, he signed on the dotted line and was inducted into the Air Force. There had been no word yet about OTS, so he reported to Lackland Air Force Base as an Airman Basic.

Three weeks into the six weeks of basic training, Gardner's instructor called him in from the physical training field and told him to pack up – he'd been accepted into OTS. "That was the only thing my agriculture degree ever did for me," he says. "It allowed me to be an officer." Being an officer meant being in a plane instead of on the ground. He was sent to California three months later for a year of navigation school.

One year in California led to three, and he had a powerful longing to get home to Texas. He bid on the two Texas choices offered to the new navigators and was assigned to Dyess Air Force Base in Abilene, flying TC 135 refueling tankers (the equivalent of Boeing 707s).

Toward the end of his four and a half years at Dyess, Gardner decided to put his G.I. Bill benefits to use; he still wanted to be a pilot. During his last year, he trained in civilian schools to get his advanced ratings, including flight instructor, and made friends at the Abilene airport. When he got out of the service, he spent a year and a half at the airport as a flight instructor, making friends as he trained new pilots and worked with the old ones. He met some corporate pilots, and eventually one of them asked if he would like to fly for an oil company. This was during the height of the oil business years, and Gardner joined as one of seven pilots, flying four planes and maintaining their own mechanic crews. He enjoyed his first corporate job, but it soon became his first taste of the vagaries of corporate flying – oil nosedived and he was downsized out. "Your professional career depends on the success of the company's business," he explains. "The economy affects jobs, and one of the first things the bean-counters look to trim is the flight department."

Knowing he was about to be laid off, Gardner signed on with Chaparral (a regional airline) as a part-time pilot. A friend, a former chief pilot at Chaparral who had become chief pilot at Texas State Credit (TSC) in Waco, needed another pilot and invited Gardner to join him there. Moving his family to Waco, he flew for TSC for several years until the company was sold.

Once again Gardner was looking for a job, and once again friends contacted him, this time with two possibilities. The Texas Farm Bureau (TFB) needed a second pilot, and a former flight student named Rick Perry (later the governor of Texas) needed a pilot in his then-position of Texas State Commissioner of Agriculture. The day of his conversation with Perry, however, the state government decided that all state planes and pilots would be pooled instead of department heads hiring their own pilots and maintaining their own planes. Pilots would be assigned from the pool as flights were scheduled, with no choice as to whom they flew with or for. Gardner and his family were happy in Waco, and he preferred working with someone he knew and trusted as a safe, skilled pilot; accordingly, he chose to work for TFB.

Now chief pilot, Gardner is in his nineteenth year at TFB. TFB has a company policy that any time there are passengers on the plane, there must be two pilots in the cockpit; Gardner always works with a second full-time pilot. To cover times when one of them has to be off or is on vacation, Gardner runs a copilot instructor program, calling on young pilots he meets in his position on the advisory board of the TSTC Waco's APT program and other pilots he knows. He hires them on a contract basis to serve as copilot for a day or several days, giving them a chance to log hours in the company's King Air 200 turboprop. Still a certified flight instructor (CFI), Gardner can teach them the plane and sign off as "training received." Then when he needs a "right-seat sub," he has several pilots on whom he can call.

Gardner strongly encourages new pilots to take every opportunity to fly. "Insurance companies want you to have 2000-3000 hours before they consider you a good risk to fly for any business they insure," he states. The general trend in corporate flying is to have two pilots, regardless of the plane (usually a twin-engine), to qualify for a discount on insurance rates for the flight department.

Gardner credits his time as a flight instructor for paving the way for his career. He recommends it to students as a way to pay bills, while simultaneously building experience and flight time. It also gave him the chance to meet a lot of people, which was his primary job-hunting tool. He called people he knew when he was laid off and asked them to call about anything they heard might be coming open. He also counsels new pilots to make a good resume and keep it current, building flight hours and experience "at every opportunity." He also advises, "Make contacts. Fly as much as you can, wherever you can. Sell yourself and get the word out."

When he looks for a new pilot, the candidate must have the minimum credentials: commercial certificate, multi-engine rating, and flight instructor credentials. It is desirable to have "time in type," he says, meaning flight hours in the specific type of plane the company owns. It is also important to have a good safety record with no history of accidents. Gardner is especially concerned about safety, and as an FAA Aviation Safety Counselor, he conducts pilot safety seminars twice a year.

Gardner also recommends that in an interview, pilot candidates ask their own questions and one in particular: "If I become your pilot, will you have a problem when I come to you and say we can't go because of problems with the weather or the condition of the plane?" The answer should be "No problem," Gardner says. "If they say, 'We have to fly at all costs,' that's a big red flag that there are going to be problems at that company." One thing he likes about his position at TFB is that he answers directly to the chief executive officer, who backs his opinion on the safety of flight conditions.

TFB pays expenses for overnight flights, and Gardner carries five company credit cards specifically for buying fuel; that way they can refuel with no problem anywhere they land. He explains that this is pretty standard for the industry, which operates under FAA Part 91, non-scheduled corporate flying and requires no strict procedural structure. Insurance requires regular proficiency training, which Gardner and his copilot take annually, both in the classroom and in simulators built to replicate the King Air 200 cockpit. According to Gardner, about 80 percent of his flying is in-state, and 100 percent of it is official business. Although any department of the TFB can contact the chief executive officer and request a flight, as chief pilot, he participates in scheduling, as well as some ancillary duties related to the flights

Corporate flying has been a good career for Gardner. Although it took time to find the right place, he finds great satisfaction in a job where he feels he fits best with both the company he works for and the people he works with.

In the past, military pilots most often were trained while they were already in the service. Now, however, it is much more common for new service members to have completed basic flight instruction and received at least a private pilot's certificate before joining. A pilot must be an officer, which requires that the candidate have a bachelor's degree, as well. New pilots are put through a ten-week screening school. At the end of the ten weeks, the students must demonstrate a high level of flying skill to be accepted into the pilot program, where they are trained on military aircraft. It is extremely difficult to achieve sufficient flying expertise in just ten weeks, so previous experience is virtually a requirement. Military pilots, because of their training and experience, are often highly desired as civilian pilots. To be a military pilot, though, one must also accept the "entire package," as Compton says: the military chain of command, advancement requirements, assigned duty stations, and military rules and regulations.

Companies hiring pilots for corporate flying, however, do not always have a preference for candidates with bachelor's degrees. They are more concerned with flying skills and the ability to deal politely and effectively with corporate executives or corporate clients, in the case of fractionals.

Fractionals are companies that sell shares of time on corporate-level aircraft, usually jets. If a company wants access to a private jet but not the expense of its own aircraft, it can buy time on a jet owned by a fractional. Typically, ten different companies will buy a share of time on the same aircraft, similar to a vacation-site timeshare arrangement. According to Ed McNeill, Department Chair of the APT program at TSTC Waco and a former fractional pilot himself, flying corporate jets for fractionals is a rewarding flight experience. Schedules are erratic, however, with flight times and lengths varying considerably, but the hours and pay can be good.

While educational requirements may vary, all employers are concerned about the character of their pilots. Whether transporting passengers or just cargo, employers look for men and women whom they feel can be counted on to perform their jobs safely and responsibly. Lack of professionalism, poor grooming, inappropriate clothing, poor communication skills, and bad attitudes will knock some candidates out of the running for jobs, or even out of one they already have. Bad attitudes include cockiness and a feeling of being entitled to a job. Job placement experts recommend paying attention to the "Four Hs:" honesty, humility (pilots with an attitude of service), humor (being easy to get along with, not being a comedian), and homework (knowing your stuff and knowing the company, too).

Physical Requirements

In addition to the myths about working conditions and pay, Viera points out that some "common myths" of commercial aviation involve physical capabilities for the pilot candidate:

> **Myth #3:** A pilot must have 20/20 vision.
> Not true, according to Viera. "As long as your eyes are correctable to 20/20 with glasses, contact lenses, laser surgery, or what have you, you can be a commercial pilot." (Military regulations may differ.)
>
> **Myth #4:** A pilot must have extra-fast reflexes.
> Again, not true. "Anyone with normal reflexes can learn to be a pilot," according to Viera.

There are some physical requirements, however. To receive a private pilot's certificate (the first step required for all advanced flight training), the student must receive at least a third-class medical certificate through an examination by an FAA-designated Aviation Medical Examiner before being allowed to fly solo in a powered aircraft. The medical examiner checks general health, eyesight, and hearing. A third-class medical allows the pilot to exercise the privileges of a recreational certificate (REC) or private pilot certificate. (REC is more restrictive than a private certificate and cannot be used in place of the private certificate to go on to further, more advanced certification.) It does not allow a pilot to fly "for compensation or hire." The certificate is valid for three years for pilots younger than forty years old or two years for pilots over forty.

Profile: *Bill McNeill, APT Instructor*

Department Chair of the APT program at TSTC Waco, Bill McNeill says he "kind of backed into flying. I knew little about airplanes until after college." Born and raised on a farm in northeast Kansas, McNeill graduated from college with a degree in agriculture. His plan to work on the family farm derailed when he received a draft notice right out of school. Opting to visit an Air Force recruiter instead of being drafted, he signed up, did well on the entrance exam, and went directly into pilot training.

Basic flight training was in a Cessna. McNeill then spent six months learning to fly a T-37 fighter jet, followed by six more months training in a T-38, another fighter jet which can reach supersonic speeds. "It was very complicated," McNeill remembers. "We did formation flying and maneuvers. They really pushed us to get through in a year."

Assigned to Psychological Warfare Operations in Vietnam, McNeill flew a DC-3 for a year, dropping leaflets and playing recorded messages through speakers. After his tour in Vietnam, McNeill went back to flying jets, this time a refueling tanker, which was the military version of a Boeing 707, in the Strategic Air Command. Flying at 250 mph and connecting to another plane in midair was "really fun," he says. "It was difficult to learn, but you got to where you could make really small movements with a great deal of control. It was a big feeling of accomplishment." Even after conquering the flight techniques required, his challenges continued, as he had to fly refueling missions at night and "in really terrible weather."

His next assignment was as a pilot for the Looking Glass program. With Looking Glass, there was always a plane airborne with a general and battle staff aboard that could run a nuclear war. They stayed up for extended periods, once flying from Indiana to Guam. Lasting about sixteen hours, the flight required them to take on fuel off the coast of Washington. The extended flying hours made it necessary for McNeill to learn to fly the receiving side of the air refueling maneuvers he had performed before as a tanker pilot.

While performing his regular duties, McNeill also attended night school and earned a master's degree. "It was something they really expected you to do to continue receiving promotions."

Retiring from the Air Force after twenty years, he kept flying as a civilian. He taught Navy pilots to fly 707s (from the same campus he teaches at now) for about a year and then flew freight for about six months. Eventually he began flying for a company contracted to do training for the Navy. McNeill and his fellow pilots "followed the fleet," flying against the Navy pilots, acting as the aggressors, and providing "bad guys" for battle exercises. Stationed on land between exercises, he and the other "bad guys" would fly out to meet the fleet wherever the ships were. Usually lasting two to three weeks, the training schedule most often had them flying along the coast of South America. He recalls one mission, however, that lasted seven weeks, starting in England and following the fleet all the way to Japan.

When the training budget was cut, the contract ended. Using networking among pilot friends to find jobs, McNeill flew various types of planes and for a variety of tasks, including freight again for a while. His recollections of that job include "a hectic schedule – long days and weird hours, especially on international flights.

He then spent several years flying Lear jets for a fractional company. Mostly daytime flights, his schedule included twelve-hour days, six on and four off. He enjoyed the job, saying he "met interesting people (like Michael Dell and Roger Clemens) and went interesting places."

McNeill says because his wife wanted him home more, he eventually accepted his current position at TSTC Waco. He enjoys the school and his students, he says. "I like the multi-tasking, the variety of jobs, people, and problems. It's never the same." He considers the TSTC program to be excellent. "This is formalized. It's in-depth, standardized training. It's the best way to do it – a civilian school." TSTC Waco has a practice of hiring its best graduates as flight instructors for the next year, he adds. "We have a really good group of instructors here, with good attitudes. You can be a super pilot and really look the part, but if you don't have a good attitude, we don't want you. We've been fortunate in hiring people who love to fly and hope to work their way up the aviation ladder."

McNeill still loves flying and hopes to pass on the passion to the next generation, encouraging students to persevere: "Anybody can make it through if they just stick with it."

The second-class medical allows a pilot to exercise the privileges of a commercial pilot certificate "for compensation or hire" for one year. Then, if not renewed, the certificate reverts to a third-class medical, restricting the pilot to the limits of the lower rating. Many flight schools that confer

degrees for professional pilots require an incoming student to possess a second-class medical before they begin classes.

A first-class medical allows the pilot to exercise the privileges of the airline transport pilot (ATP) certificate for six months. If not renewed every six months, it reverts to a second-class medical, then to a third-class medical with the accompanying restrictions.

Medical checks continue throughout a pilot's career. Only FAA-approved physicians can conduct a flight physical and issue the medical certificate. Approved physicians in a particular area can be located through the FAA.gov website. Once the physical is passed, the doctor completes the certification paperwork and files it with the FAA.

Profile: *John Foster, Corporate Chief Pilot and FAA Designated Pilot Examiner*

John Foster acquired a love of airplanes while helping out at the local airport and learning to work on airplane engines with the owner/ mechanic. When he accepted an offer of a flight one day, he learned to love flying as well. Soon he was working all week to pay for an hour of flight training, and eleven months later he earned his private pilot's certificate at seventeen.

At the age of nineteen, he started APT at TSTC Waco. Foster says he "did not really see a point to high school," and as a result didn't study. However, in APT, he worked hard on his education. He loved his classes, loved the place, and made friends he still sees flying around the country. He spent some time flight instructing at TSTC and then became chief flight instructor at the McGregor Airport a few miles down the road. He later spent six months flying for Lone Star Airlines (a regional line out of Dallas). "But I didn't care for the on-the-road lifestyle," he says.

He went to work as the chief pilot for Brazos Air, an air charter company with three leased airplanes. Foster found that he enjoyed corporate flying, and the salary was much better than the airline. "That [the airline] was the least amount of money I ever made." Foster cautions, "A lot of young airline pilots can qualify for food stamps. You have to love flying and persevere through it."

Two years later, in August of 1999, he was hired at BU as a copilot, flying four different planes in his six years there: a King Air, Beechjet, Cessna Citation, and another King Air. With "more pay and less work," he enjoyed his time at BU so much that he thought he would never leave there.

Then, in December 2005, Foster got an offer from Home State Insurance to become chief pilot. He has found the company to be very stable and family-oriented, which appeals to him. The flying schedule suits him as well. During an average month, he might be gone on one five-day trip, but the rest of the month he flies only day trips, giving him plenty of time to spend with his wife and two young boys. "That's not standard in corporate flying," Foster says. "With most places you're gone a lot. I found the right place, the right fit for me." He also enjoys the places he goes. "It may be anywhere in the United States and sometimes to the Caribbean for conventions, training seminars, or business-to-business meetings."

Foster's flight department consists of himself, two part-time copilots, a Cessna Citation jet, and a Cessna 182. All the corporate flying is done in the Citation, with the 182 used primarily as a "chase plane" to follow the jet when it is taken in for maintenance and to bring the pilots home. Currently the copilot position is contract and part-time. Foster's boss is also a pilot, and "whenever he's on the plane, he's the copilot." For the second copilot, he looks for someone with at least a year of experience as a flight instructor. He also looks for someone with a commercial certificate, multi-engine and instrument ratings, and preferably a multi-engine instructor.

Most of his copilots have had no experience in jets, so Foster teaches them the aircraft. "It's a paid education," he says. His previous copilots have gone on to full-time corporate jobs, partly due to the experience they gained with him. Besides the minimum skills, Foster says he looks for someone who is calm, personable, attentive to the needs of the passengers, and helpful loading the bags. "A team player," he says.

Foster gets to know a lot of young student pilots and observes their skills and personalities in his position as an FAA Designated Pilot Examiner. He keeps an eye out for young people with good flying skills as well as good people skills. When Foster is not flying for Home State Insurance, he tests about one hundred student pilots a year. Recommended to the FAA when the previous examiner retired, Foster had already met most of the requirements, but he went to Oklahoma City for training and tests, both written and practical. Once licensed, he says, an examiner can keep the job until he or she chooses to leave. Most examiners are retired pilots, and of the forty in the last required continuing education course he took, two were women.

Foster describes corporate flying as "the road less traveled." The corporate career path tends to be "kind of hit-or-miss," and most corporate jobs are found by networking and word-of-mouth. "I don't know anybody who got a job by answering an ad," he says. "It may have been advertised, but it's usually a personal recommendation that gets you in." He also says that a four-year degree is not nearly as important in corporate flying as it is for airlines, but that a two-year technical degree is very beneficial. He also counsels patience.

"It takes a while to find the right place. Once you find it, you want to stick with it." He feels he has definitely found his place at

Home State Insurance. "I plan to retire from this one. I enjoy running all the aspects of the operation: scheduling, maintenance and inspections, working with the caterers, and sweeping the hangar. It took perseverance, but after twelve years of hard work, I'm an overnight success!"

Except for in the military, it is usually the pilot's responsibility to arrange and pay for flight physicals. Insurance typically does not cover the cost, but some companies will cover it for their employees. A employer may also require periodic drug screens. Certain physical conditions can cause the rescinding of the medical certification, if they are deemed to be potentially dangerous or damaging to the pilot's ability to properly control an airplane.

Tasks/Skills

According to O*Net Online, the following are the most important, or core, tasks and skills required for airline and commercial pilots, and the most important supplemental skills:

Core tasks/skills

- Use instrumentation to guide flights when visibility is poor
- Respond to and report in-flight emergencies and malfunctions
- Work as part of a flight team with other crew members, especially during takeoffs and landings
- Contact control towers for takeoff clearances, arrival instructions, and other information using radio equipment
- Steer aircraft along planned routes, using autopilot and flight management computers
- Monitor gauges, warning devices, and control

panels to verify aircraft performance and to regulate engine speed

- Start engines, operate controls, and pilot airplanes to transport passengers, mail, or freight, adhering to flight plans, regulations, and procedures
- Inspect aircraft for defects and malfunctions, according to pre-flight checklists
- Check passenger and cargo distributions and fuel amounts to ensure that weight and balance specifications are met
- Monitor engine operation, fuel consumption, and functionality of aircraft systems during flights
- Confer with flight dispatchers and weather forecasters to keep abreast of flight conditions
- Coordinate flight activities with ground crews and air traffic control and inform crew members of flight and test procedures
- Order changes in fuel supplies, loads, routes, or schedules to ensure safety of flights
- Choose routes, altitudes, and speeds that will provide the fastest, safest, and smoothest flights
- Direct activities of aircraft crews during flights
- Brief crews about flight details, such as destinations, duties, and responsibilities
- Record in log books information such as flight times, distances flown, and fuel consumption
- Make announcements regarding flights using public address systems

Supplemental tasks/skills

- File instrument flight plans with air traffic control to ensure that flights are coordinated with other air traffic
- Perform minor maintenance work or arrange for major maintenance
- Instruct other pilots and student pilots in aircraft operations and the principles of flight

- Conduct in-flight tests and evaluations at specified altitudes and in all types of weather to determine the receptivity and other characteristics of equipment and systems
- Evaluate other pilots or pilot-certificate applicants for proficiency
- Load smaller aircraft, handle passenger luggage, and supervise refueling
- Plan and formulate flight activities and test schedules and prepare flight evaluation reports

Conclusion

Although there will be periodic flat spots and economic slowdowns affecting the number of professional aviation job openings, the industry as a whole is expected to continue growing, with the majority of the growth occurring in smaller regional airlines. It is a demanding field. Skills are constantly checked, and safety is monitored closely by the FAA. People skills are important, as well as flying skills, in order to relate successfully both with passengers and fellow crew members and to complete the tasks required on each individual flight. Technology in the field of aviation continues to advance, with new aircraft being designed, and new instruments developed.

New skills and certifications will continue to be required in order to advance in experience and be qualified to pilot the most sophisticated and complicated new aircraft. If you love flying, however, these are all opportunities to do more of it.

Notes

CHAPTER TWO

Aviation Pilot Education and Certification

"Knowledge is the wing with which we fly to Heaven."
– William Shakespeare

Educational Requirements

Anyone wishing to pilot an aircraft of any sort must be certified by the Federal Aviation Administration (FAA) to fly. The most common, basic level of certification (and the only one which can be used to continue on to professional pilot status) is the private pilot's certificate with a single-engine certificate. This basic level often can be obtained at local airports or airstrips, though these private schools cannot confer degrees of any kind. All pilots who wish to fly professionally must have this first level of certification accomplished. To become a commercial pilot, however, much more training and more advanced certifications are required, such as a two-year associate degree at technical schools and a four-year bachelor's.

In addition to the required certifications, a minimum number of flight hours must be accumulated to qualify a pilot for different levels of commercial flying. This can be an expensive process, as planes must be rented and fuel purchased for all of these hours. A pilot must have logged 1500 flight hours for a position at a regional airline. These are short-distance, smaller-plane airlines, and often subsidiaries of larger airlines. For example, American Eagle is a regional subsidiary of American Airlines. Ed Viera, Chief Flight Instructor at Texas State Technical College (TSTC) Waco, describes this system as working much like the minor and major leagues in baseball. A player hired right out of college plays first for a minor-league team. When he has acquired the needed experience and honed his skills at that level, he moves up to playing in the major leagues.

Realistically, in order to advance to a career as a professional pilot for a major airline, a bachelor's degree is almost a necessity. Many pilots apply for jobs every year, and having a bachelor's degree makes a candidate more marketable, particularly above a candidate with an associate degree.

Technical schools, however, can offer some valuable advantages to achieving both the commercial pilot certification and a bachelor's degree. For example, TSTC Waco offers an associate degree in five semesters. In addition to more in-depth training and education than is possible in a small private school, achieving the associate degree allows graduates to work in the industry while completing the remaining requirements for a bachelor's degree either on campus or online. TSTC's program in particular allows students to acquire several flight instructor certifications in basic flight, instrument flight, and multi–engine aircraft.

Every year, the TSTC program hires several of its own best graduates to work as flight instructors for the next year or two. Instead of paying to rent planes and buy fuel, these former students build up the required flight hours while being paid for doing it. They can also work towards their bachelor's degree (which can be in any major) at the same time. The additional training to become a flight instructor and the hours spent training new students increases a pilot's knowledge and understanding of aircraft systems and his or her skill in recognizing and handling possible in-flight problems. According to Bill McNeill, department chair of the Aircraft Pilot Training (APT) program at TSTC Waco, the increased skills and experience make a person a much better and more reliable pilot. "Airlines know this very well," he says, "and they value it very highly in looking at job applicants."

Classroom and In-Flight Instruction

Many courses for flight training, like navigation and meteorology, are conducted in classrooms. Regular tests are given to check each student's grasp of the material. This is where many students "self-eliminate," observes Viera. Some younger student pilots are too interested in partying in their first experience away from parental control. TSTC Waco Flight Instructor Jeff Benavides agrees. "The best students are a little more mature. You have to have the proper mindset, and you can't cut class. Courses are very concentrated in the technical school and move quickly. You are accountable for everything in class. It all shows up in the plane later." Because pilot certification is federally controlled, there is no latitude in the testing. A student cannot fake his or her way through a course; the required knowledge and skills must be developed to pass the tests.

In-flight instruction is conducted in stages, and each flight certification has three stages. After a classroom introduction to information and possibly practice in a flight simulator, the student pilot flies with an instructor to develop and practice the necessary skills. In order to move on to the next stage, the student must pass a written test, complete a minimum number of flight hours, and receive a sign-off from the flight instructor indicating he or she believes the student is ready to progress to the next stage. At that point, the student must pass a check flight with a designated "check airman" to confirm the student's skills.

When all three stages have been completed successfully, the student must pass a written test and a check flight conducted either by the FAA (which are quite rigorous and often difficult to schedule due to the demands on the FAA's time) or by an FAA Designated Flight Examiner, who is an experienced pilot certified by the FAA to conduct check flights for student pilots. The examiner does not work for the school and is accountable to the FAA for thoroughly testing the student pilot. He or she flies with the student and watches for a prescribed level of knowledge and skill. If the student pilot successfully completes the check flight and passes the flight tests, he or she is granted the certification level and may then move on to the next level of flight instruction and certification. Each stage must be successfully completed in order to progress to the next.

Be aware beforehand that various factors can interfere with the completion of the flight hours required for each level. Bad weather can delay flight for some time, airplane maintenance concerns and requirements may take a plane out of action for a while (possibly for several weeks or months if an accident has seriously damaged the aircraft), or

planes may be unavailable because of other commitments. It is almost impossible to guarantee that any particular student will graduate in exactly the time frame spelled out in the college catalog. If you are focused, determined, and stick with the training, however, you can accomplish your Associate of Applied Science (AAS) and be on your way to becoming a professional pilot.

Some flight schools offer an AAS certificate of completion. This is a one-year technology/academic program that confers recognition of a more limited range of aviation courses and most likely only one or two FAA certifications/ratings. This type of program is less deep and results in fewer FAA certifications, but it can possibly be used to a student's advantage in working his or her way through school. You should study the opportunities at different institutions carefully and choose the program likely to give you the best outcome for your individual goals. In addition, make sure you understand whether any certificate mentioned refers to an FAA certificate or an AAS certificate of completion.

Most airline pilots have a bachelor's degree in addition to their pilot certificates. This means that to compete effectively in the job market at that industry level, you will most likely need to have a four-year degree as well. Some college majors may make additional career paths possible as well, including aircraft design and testing and airline/airport management. There are also several paths a student can take to accomplish a bachelor's degree, including getting an AAS first and then completing the remaining two years either on campus, at night classes, or online.

Profile: *Angel Newhart, Professional Pilot*

Angel Newhart has wanted to be a pilot since she was six years old. She says she cannot remember for sure, but she thinks her family was at an air show when she made her decision. "I remember seeing them (the planes) and getting to touch and sit in one, and I just knew. Ever since then, I've wanted to be a pilot."

Army service was the first stepping-stone toward her goal. She had hoped to become a pilot in the military, but some obstacles got in the way. She served instead as an OT-1D Mohawk navigator and aerial observer. After the army, she learned to fly at TSTC Waco, taking three years to get through the program. She then spent a year at Tarleton State University in Stephenville, Texas, where she received her bachelor's degree. Afterwards, she returned to TSTC and worked as a flight instructor for about three years, building up hours and experience.

Her next stop was as a first officer (FO) for Colgan Airlines, where she worked for almost two years. Life again interfered with her plans, and she went back to TSTC Waco as a flight instructor. Newhart focuses on the positive elements of the change, though. "I love to teach," she says, "and I like flight instructing." She also serves as the program's check airman, doing the in-house stage check rides.

Besides her own love for flying, she believes aviation holds tremendous opportunities for women. Almost every year, she travels to Northwestern University in Fort Worth for their "Expanding Your Horizons" discovery day for sixth-, seventh-, and eighth-grade girls. Representatives from all areas of aviation and other fields come to talk to the students to encourage them to study math and science and illustrate the doors that those subjects can open for them. Newhart tells of one girl who told her, "Well, I could be a flight attendant, but they're not going to hire me to fly." Newhart's answer was, "Yes, they will! If you have the training, you're just as qualified as the guy next to you." It is important to get the word out, she says. "We have just as many opportunities as the guys do now. There's no more 'boys' club.' The entire time I've been with the military and the civilian side of aviation, I have never been treated inappropriately or been held back for promotion or anything. I love flying, and I've had nothing but good experiences."

Although Newhart has taught at both Part 61 and Part 141 flight schools, she particularly enjoys the group dynamics and cooperation available at the Part 141 schools. "If you have a question or you just want to throw something off of somebody, you just walk over to the desk and say 'Hey, I've got a question,' and any one of us will answer. In fact, the instructors encourage their students, if they are having trouble with something, to get another point of view."

Newhart still has many goals in aviation. She has been preparing to go back into the Army to learn to fly helicopters. After she qualifies as a rotary pilot, she intends to return to TSTC Waco as a flight instructor for rotary aircraft, while also flying them for the Army Reserves.

Her best advice to students wanting to become professional pilots is to be self-motivated. "Everybody has trouble finding financial solutions to pay for flight fees. Everybody has to go through the classes and study to get a high GPA. Everybody needs to have a minimum amount of skill. Your instructors will motivate you as much as they possibly can. If you're self-motivated, you can [in spite of the difficulties] push yourself through."

As far as she is concerned, "You can't find a better office. The challenge every day is different, whether it be different lessons that I'm doing, different airplanes, different students, or a Discovery Flight for someone that's getting to fly in an airplane for the first time. The other day, I had an eighty-eight-year-old lady – she was the sweetest thing – from Baylor, who had never been in an airplane, and it was just amazing to get to take somebody up and show them that. I love it, and I would absolutely do it again. It was worth everything I spent on my flight training."

Corporate and cargo pilots, on the other hand, do not always need to have a four-year degree to land jobs. Some corporations do require a degree, but some require only the proper certificates and experience. For either situation, beginning with a two-year associate degree program and either going on to complete a bachelor's or getting into commercial piloting without an advanced degree can be a very cost-effective way to achieve the necessary training, ratings, and certifications.

For reference, O*Net Online has published this Distribution of Educational Attainment Chart:

Occupation	Percent of employees aged 25 – 44 in the occupation at left whose highest level of education is:		
	High school or less	Some college	Bachelor's degree or more
Airline pilots, copilots, and flight engineers	5.9%	20.5%	73%
Air Transportation workers	9.4%	37.6%	53%
Transportation and material moving	58.6%	31%	10.4%
Total, all occupations	38.8%	29.1%	32.1%

Certification Requirements

Most certifications can be earned in various types of flight schools or with individual lessons from certified instructors. Certifications are built up one upon another and must be earned "from the bottom up," meaning that a private pilot certificate must be achieved before a commercial certificate, and a commercial certificate must be achieved before an airline transport certificate.

In addition, each level of flight training is broken up into three sections, or stages. Before proceeding to Stage Two, student pilots must complete Stage One and pass a check flight, in which a Check Airman flies with the student and reviews the skills he or she is required to have mastered at that point. Stage Two must be completed and the check flight passed before moving on to Stage Three. The flight instructor must sign off on the completed flight training, an FAA written exam must be passed, and a flight test administered by a designated FAA flight examiner must be successfully completed in order for a student pilot to move on to the next level of training and certification. This system of stages continues in each new certification level for each flight class.

After achieving a basic certification, such as private pilot or commercial pilot, a student can acquire various "ratings." Ratings represent FAA approval for different types of flying within a base certification. For example, after becoming a private pilot, a student can train for instrument flying or for flying a seaplane, a tailwheel-configured aircraft, a helicopter, a multi-engine plane, or aerobatics. Beginning in 2006, the FAA requires that for two-pilot aircraft, the copilot (sometimes called FO) must acquire an second-in-command (SIC) rating for the type of plane they fly.

Student Pilots

Not a formal certification level, "student pilot" designates a pilot who is training for a private pilot certificate, either before or after the first solo. A student must obtain a third-class medical certificate through an examination by an FAA-designated Aviation Medical Examiner before being allowed to fly solo in a powered aircraft. (The medical certificate for a student pilot has a "student certificate" printed on the back.) Training with an instructor may begin at any age, but a student learning to fly an airplane cannot fly solo until age sixteen.

These are the most common levels of certificates received in the process of professional pilot training:

Private Pilot

The private pilot certificate allows a pilot to fly passengers for personal transportation and business. The pilot must be at least seventeen years old (sixteen years old for a glider or hot air balloon), have a minimum of forty hours of flight experience and training (thirty-five hours if the school is under Part 141), and pass at least a third-class medical exam, a written exam, and a flight test with a designated FAA examiner. A private pilot may not "fly for hire or compensation," but may share expenses equally with passengers.

Certified Flight Instructor

A pilot holding a private pilot certificate and an instrument rating (an instrument rating allows a pilot to fly under Instrument Flight Rules [IFR] in conditions that make it impossible to see the ground) and who passes two written tests and a practical flight exam is FAA-rated to give flight instruction at a specific level of certification. The flight instructor rating is specific as to the type of instruction authorized, such as single-engine airplane certified flight instructor (CFI), multi-engine instructor (MEI), certified flight instructor instrument (CFII), and helicopter. A private pilot is not required to be certified as an instructor in order to earn higher certifications, but employers are aware of the extra training, experience, and capabilities that earning an instructor certification requires. Having one or more instructor certifications increases your chances of being hired.

Commercial Pilot

To achieve an FAA commercial pilot certificate, a pilot must be at least eighteen, with a private pilot certificate and a minimum of 250 flight hours completed, or logged, along with various other sub-requirements. He or she must have an instrument rating and must pass a commercial written test and a commercial flight test with an FAA examiner. This certificate allows a pilot to fly for compensation or hire, often in a wide variety of commercial general aviation operations including sightseeing, crop dusting, glider towing, and flight instruction.

Airline Transport Pilot

An ATP certificate requires a minimum age of twenty-three, with both a private and a commercial certificate, an instrument rating, at least 1500 hours of flight experience, and passing an ATP written exam and a flight test with an FAA examiner. It is not usually available in standard flight schools. Training may be provided by the employer after hiring a new pilot, or it can be obtained in special ATP flight schools (typically found at or near large airports). This certification is mandatory for captains of major scheduled airlines, regional carriers, scheduled commuter airlines, and some commuter and on-demand operations. It is also a hiring requirement for many pilot positions in corporate and commercial general aviation flying.

Multi-Engine Pilot

Private Pilot and Commercial Pilot training qualify a pilot to fly single-engine airplanes only. In order to progress to an airline transport pilot or many other types of flying, a commercial pilot must acquire a "rating" for a multi-engine airplane. The training procedure is the same as for the first two levels.

Training usually continues throughout the career of an airline pilot, with major airlines especially requiring regular sessions in simulators to train on various trouble scenarios. Airlines will also train pilots on new aircraft they may be assigned to. FAA-designated examiners make periodic checks of a pilot's skills, and medical flight exams must be passed on a regular basis, with the timing depending on the level of the certificate. These checks must be passed in order to retain the pilot's certification level.

The Two-Year Associate of Applied Science Degree

Many technical colleges across the country offer two-year AAS degrees in aviation. While they may differ in the specific classes offered, a two-year AAS degree will most often allow the student pilot to achieve several certifications and ratings, beginning with the private pilot certificate and possibly including a commercial single-engine certificate, multi-engine commercial rating, instrument flight rating, and one or more instructor certifications: CFI, CFII, and MEI. These degrees also include some core non-aviation academic classes and electives.

It is important to research the different schools to find out the advantages and disadvantages of each one you are interested in. One of the differences you may find is that some schools own and maintain their own fleet of airplanes for training the different levels of certification. Others have contracts with local flight schools at regional airports to provide the flight training for their program. This may affect the cost of the flight hours and how the payment is made, whether to the college or to the flight school.

Here are degree plans for two-year degrees from several different colleges:

Central Texas College
6200 W. Central Texas Expy.
Killeen, TX 76549
254.526.1241

The Central Texas College Aviation Science Department offers a two-year program leading to an AAS degree and the commercial pilot certificate with an instrument airplane rating. The curriculum includes fourteen core aviation classes, which are transferable to upper level institutions that offer bachelor's degrees in aviation. The Central Texas College Flight Training is conducted under FAA regulations for Federal Aviation Regulations (FAR) 141 Approved Schools. The local area offers an excellent training environment, which allows for efficient use of flight time. Central Texas College also offers advanced flight training for flight instructors and for multi-engine ratings.

Extra curriculum activities include Alpha ETA Rho, the National Intercollegiate Aviation Fraternity, and the Central Texas College Precision Flight Team, which competes in the National Intercollegiate Flying Association competitions. Central Texas College is a member of the University Aviation Association.

	Aviation Science	
	Associate of Applied Science Degree	Total Hours: 72
First Semester		**Credit Hours**
AIRP 1315	Private Flight	3
AIRP 1417	Private Pilot Ground School	4
AIRP 1305	Aircraft Science	3
ENGL 1301	Composition I	3
MATH 1414	College Algebra	4
	Semester Total	**17**
Second Semester		
AIRP 1255	Intermediate Flight	2
AIRP 1341	Advanced Air Navigation	3
AIRP 1307	Aviation Meteorology	3
SPCH 1315	Public Speaking	3
ENGL 1302	Composition II	3
ELCT	Approved Elective	3
	Semester Total	**17**
Third Semester (Summer)		
AIRP 2250	Instrument Flight	2
AIRP 1451	Instrument Ground School	4
ELCT	Science Selection (with Lab)	4
KINE	Physical Activity Course	1
	Semester Total	**11**

Fourth Semester		
AIRP 2250	Commercial Flight	2
AIRP 2337	Commercial Ground School	3
AIRP 1343	Aerodynamics	3
ELECT	Computer Technology Skills	4
ELECT	Fine Arts Selection	3
	Semester Total	**15**
Fifth Semester		
AIRP 2355	Propulsion Systems	3
GOVT 2301	American Government I	3
AVIM 2331	Airline Management	3
ELCT	Humanities Elective (Literature)	3
	Semester Total	**12**

Central Texas College also offers a Certificate of Completion, which can be completed in two years.

	Aviation Science	
	Certificate of Completion	**Total Hours: 38**
First Semester		**Credit Hours**
AIRP 1315	Private Flight	3
AIRP 1417	Private Pilot Ground School	4
AIRP 1305	Aircraft Science	3
	Semester Total	**10**
Second Semester		
AIRP 1255	Intermediate Flight	2
AIRP 1341	Advanced Air Navigation	3
AIRP 1307	Aviation Meteorology	3
AIRP 1343	Aerodynamics	3
	Semester Total	**11**
Third Semester		
AIRP 2250	Instrument Flight	2
AIRP 1451	Instrument Ground School	4

AIRP 2355	Propulsion	3
	Semester Total	**9**
Fourth Semester		
AIRP 2239	Commercial Flight	2
AIRP 2337	Commercial Ground School	3
AVIM 2331	Airline Management	3
	Semester Total	**8**

Palo Alto College
1400 W. Villaret Blvd.
San Antonio, TX 78224
210.486.3000

The Professional Pilot Program prepares students who look forward to careers as air carrier pilots, general aviation pilots, and military pilots. Each student successfully completing the degree program will have obtained a minimum of a commercial pilot certificate with an instrument rating.

	Professional Pilot	
	Associate of Applied Science Degree	**Total Hours: 60**
First Semester		**Credit Hours**
ENGL 1301	Composition I	3
GOVT 2305	Federal Government	3
AIRP 1313	Introduction to Aviation	3
AIRP 1317	Private Pilot Ground School	3
AIRP 1315	Private Pilot Flight	3
	Semester Total	**15**
Second Semester		
SPCH	Communication Core	3
ITSC 1309	Integrated Software Applications	3
MATH	Mathematics Core	3
AIRP 1345	Aviation Safety	3
	Semester Total	**12**

Summer Semester		
AIRP 1351	Instrument Ground School	3
AIRP 2350	Instrument Flight (Lab)	3
	Semester Total	**6**
Third Semester		
AIRP 2337	Commercial Ground School	3
AIRP 1343	Aerodynamics	3
AIRP 1355	Intermediate Flight	3
ELECT	Humanities or Visual and Performing Arts Elective	3
	Semester Total	**12**
Fourth Semester		
AIRP 2349	Instructor Ground School	3
AIRP 1307	Aviation Meteorology	3
AIRP 2333	Aircraft Systems	3
AIRP 2339	Commercial Flight	3
AIRP 2336	Certified Flight Instructor - Airplane	3
	Semester Total	**15**

TSTC Waco
3801 Campus Dr.
Waco, TX 76705
800.792.8784

The APT Technology program at TSTC guides students through the challenge of gaining skills, experience, and various flight certificates needed to become a successful pilot. The five-semester flight program offers an AAS degree in Aviation. In the APT program, expect total immersion in flight activities, from classroom studies to assignments in the sky.

Students must pass the TSTC entrance exams or complete required remediation before enrolling in the APT program. In addition, all students enrolling in the Aircraft Pilot Training program must have a second-class medical in their possession at the time of enrollment.

	Aircraft Pilot Technology	
	Associate of Applied Science Degree	Total Hours: 72
First Semester		**Credit Hours**
TECH 1100	Tech Success	–
CTEX 10XX	Tech Success Seminars (3 as assigned)	4
AIRP 1215	Private Flight	2
AIRP 1301	Air Navigation	3
AIRP 1307	Aviation Meteorology	3
	Semester Total	**12**
Second Semester		
AIRP 1451	Instrument Ground School	4
AIRP 2250	Instrument Flight	2
AIRP 2331	Advanced Meteorology	3
AIRP 1313	Introduction to Aviation	3
MATH 1314	College Algebra	3
	Semester Total	**15**
Third Semester		
AIRP 1341	Advanced Air Navigation	3
AIRP 1255	Intermediate Flight	2
AIRP 1343	Aerodynamics	3
AIRP 2333	Aircraft Systems	3
ENGL 1301	Composition I	3
	Semester Total	**14**

Fourth Semester		
AIRP 2337	Commercial Ground School	3
AIRP 2239	Commercial Flight	2
AIRP 1345	Aviation Safety	3
AIRP 2355	Propulsion Systems	3
ACMG X3XX	General Ed Social Science Course	3
	Semester Total	**14**
Fifth Semester		
AIRP 2349	Instructor Ground School	3
AIRP 2236	Certified Flight Instructor - Airplane	2
AIRP 2357	Turbine Aircraft Systems Ground School	3
AVIM 2337	Aviation Law	3
ACGM X3XX	General Ed Elective	3
ACGM X3XX	General Ed Elective	3
	Semester Total	**17**

In addition to ratings required for the associate degree, pilots may earn the following certificates and ratings required for aviation careers: CFI, multi-engine pilot, and MEI.

The Four-Year Bachelor's Degree

A four-year bachelor's degree is recommended for all students who wish to pursue employment with passenger airlines. The degree can be in almost any field. Some experts recommend having an educational background in an area other than aviation, based on the theory that it may help provide employment in other areas during slow economic times when airlines may layoff or furlough pilots until the aviation business picks up again. On the other hand, a degree in aviation science, engineering, or maintenance can open up other avenues in aviation-related careers, for instance airport management, air safety, logistics, aircraft design, and testing.

One of the options for obtaining a bachelor's degree is to acquire initial pilot training from a two-year academic or technical school and then enroll in a four-year institution, transferring the credits from the two-year institution. Dr. Jim Fullingim of Texas A&M – Central Texas (TAMU-CT) is a firm believer in the value of the two-year associate degree as the first step in an aviation education. The concentrated flight schedule leads to more flying time sooner, he says, and makes it easier for the new pilot to make a living after graduation and help fund the education for a bachelor's degree.

"The more advanced degree is now virtually a requirement," he adds. "It's a testament that you put in the dedication and the work to accomplish it. Because so many other pilot candidates have at least an associate degree, you make yourself more marketable, and you can compete on the same level as other job applicants. Especially when the economy gets rough – the airlines get pickier about whom they hire. The more pieces of paper you have, the better."

Fullingim has been a professional pilot himself, in his case getting a bachelor's in fine arts before he decided to pursue flying. He attended a two-year technical college to get an associate degree and later pursued his master's and doctorate. "Of all the degrees I've received," he says, "that associate degree was the most important. It changed my life."

TAMU-CT is unique, Fullingim says, in that it is the only state-assisted school offering an aviation curriculum in Texas, and it offers no lower-level courses. The program operates specifically to help students with an associate degree, or two years of undergraduate classes, to move up to a bachelor's. It provides only junior- and senior-level classes, along with graduate courses. All freshman and sophomore credit hours are transferred in. Only specific schools are assured that all credit hours will transfer, though, so research is necessary to plan your education accordingly.

Many schools offer online degree programs, allowing students to earn a bachelor's degree or higher while working as a flight instructor or in some other job. There are also colleges and universities that offer four-year programs in aviation, which include obtaining professional pilot certificates as part of the curriculum. Most universities that offer four-year professional pilot degrees do not have their own planes, so they often have agreements with local flight schools to handle the flight training. This will require students to coordinate the university academic schedule with the flight-training schedule, as well as have the ability to travel.

Profile: *Tim Compton, Aviation Sciences Instructor*

Tim Compton, Assistant Department Director of Aviation Sciences for Baylor University (BU), grew up fascinated with airplanes. "My family used to drive out to the airport, park next to the end of the runway, and watch the planes taking off and landing. Then, on a trip to Maryland, we saw F-14s flying. I knew I wanted to go into the Navy and do that."

Earning his bachelor's degree in aeronautical science from St. Louis University in 2000, Compton became a naval officer in 2001, flying for the Navy until 2005. He continued to be fascinated by the challenges of flying – the precision of aviation and aeronautics. "It's constantly changing," he says, "whether it's on a single flight and how the weather changes from point to point or how technology changes."

After his Navy experience, he went back to school to pursue a graduate degree, but it was important to him to remain in aviation. He was offered a graduate assistantship in the aviation sciences program at BU and began teaching in the department. After finishing his Master of Science degree in 2006, he was asked to stay on as a lecturer. In 2008, he became the assistant director of the program.

"As a college student," Compton says, "working on my degree was the first step toward pursuing my goal. I had a dream to fly jets for the Navy, and I could not have reached my goal and my dream had it not been for the instructors who encouraged me, mentored me, and kicked my butt when I needed it. Those instructors provided me the opportunity to chase and to capture my dream, and I just feel 100 percent compelled to be in aviation education so I can provide that for students who are coming up in the ranks and who want to pursue their dreams. I feel a calling for giving back to the field of aviation, and specifically academic aviation."

Compton believes in the value of a four-year degree, averring that it opens doors a two-year degree might not. "For instance," he explains, "you have to have a four-year degree to put in an application for military aviation. You cannot become an officer in the military without a four-year education." He admits, though, in the area of commercial aviation, a bachelor's degree is not necessarily required. "We work hand-in-hand with airline recruiters, and they say that while it is not required, when there are so many applications going in for so few jobs, you have to quantify the qualifications. You have to somehow pick the best. On paper, you can't quantify flying skills, so a four-year degree is more sought after. It certainly propels an application to the top of the stack."

Compton says he makes a point of keeping his CFI certificate active and valid. "Every semester I fly with at least one student, and I get occasional chances to fly with some of the local companies. I stay as active as possible. I think it's incredibly important to bring recent, real-world scenarios into play in the classroom," he states. In addition, he keeps sharpening his flying skills. "This summer I'll be up in Alaska, flying goods and people back into the small Inuit territories during the couple of months of good flying weather, and I'll be able to bring those experiences back to the classroom."

Another area of aviation where Compton has found great satisfaction is in research. His master's thesis explored alternative fuels for aircraft, and he continues to work on various research projects. He is a member of the Society of Experimental Test Pilots and a requested speaker for area Boy Scout troops on aviation-related topics, taking aviation to the next generation whenever and wherever possible.

One advantage of attending a college or university is that you can earn college credit at the same time as your flight training, getting your pilot certificates and ratings along with the academic curriculum. Upon graduation, you might have earned an associate and/or bachelor's degree(s) in an aviation/aerospace-related career field. Another advantage is the various financial aid opportunities available to eligible students.

Some institutions also have formed partnerships with various regional and national airlines, incorporating the airline's training program into their flight-training curriculum and allowing a student to work an internship with the airline. The goal of these relationships is to create opportunities for students to become new-hire pilots for the partnering airline. Dr. Fullingim points to several semester-long internship opportunities TAMU-CT offers with American Airlines, which guarantee a successful intern an interview with the company after graduation. However, although a regional airline internship or training program may guarantee you an interview, it does not guarantee you a job.

A couple of disadvantages of attending a college/university are the length of the training program and costs. At some institutions, pilot certificates and ratings are generally earned over a four-year time frame. Others may allow you to earn all of your certificates and ratings within two years. It is important to review the school's flight training program and curriculum, and choose the direction that will be most advantageous to you.

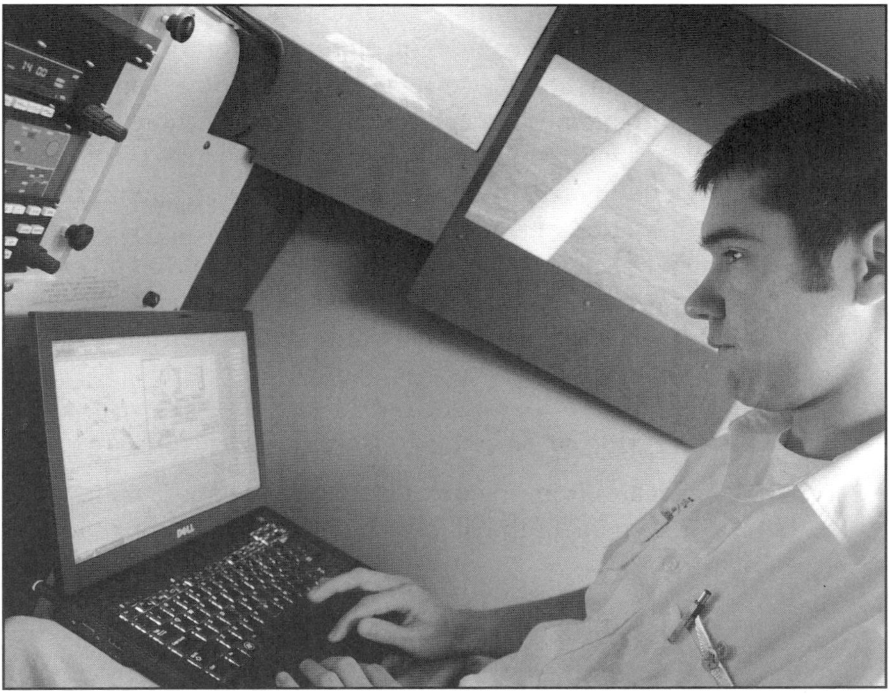

Here are degree plans for four-year degrees from several different colleges:

LeTourneau University
2100 S. Mobberly Ave.
Longview, TX 75607
903.233.3815

LeTourneau University offers different career emphases to choose from within the Professional Flight Degree program. The Bachelor of Science degree in Professional Flight is designed for the student preparing for a career as a professional pilot. This major leads to certification as a commercial pilot, multi-engine, and single engine with instrument ratings, with a strong emphasis on large commercial aircraft operations. Also offered are a management concentration, a marketing concentration, and a military aviation concentration.

LeTourneau also offers an Aeronautical Science Degree, allowing a student to combine flight training with the aircraft systems science curriculum, leading to a commercial pilot multi-engine and single engine with instrument ratings as well as the Airframe and Powerplant mechanic certificate. This is a highly technical degree with a strong emphasis in general aviation operations including corporate aviation. The Bachelor of Science in Professional Flight can be completed in eight regular semesters and one summer session. Within this degree program, the school also offers a mission aviation concentration.

Bachelor of Science in Professional Flight

Graduates from the program will have approximately 190 flight hours in the school's planes. Certifications earned include:

- Private Pilot Certificate and Instrument Rating
- Commercial Pilot Certificate
- Multi Engine
- Single Engine
- Instrument Rating

Optional flight instructor certifications:

- Multi Engine Airplane
- Single Engine Airplane
- Instrument Rating

	Bachelor of Science in Professional Flight	Total Hours: 126
First Semester		**Hours**
AVTC 1401	DC Electricity Laboratory	1
AVTC 1403	DC Electricity	3
ENGL 1013	English Composition I	3
BIBL 1033	Biblical Literature	3
KINE 1512	Concepts of Lifetime Fitness	2
LETU 1101	Cornerstones of Life & Learning	1
MATH 1203	College Algebra with Applications	3

	Fulfill English Proficiency Requirement	0
	Semester Total	**16**
Second Semester		
AERF 1113	Flight Science I (Private Ground School)	3
AVTC 1001	Foundational Concepts of Aviation Studies Laboratory	1
AVTC 1003	Foundational Concepts of Aviation Studies	3
COMM 1113	Introduction to Speech Communication	3
ENGL 1023	English Composition II	3
MATH 1252	Trigonometry	2
	Semester Total	**15**
Third Semester		
AERF 1111	Primary Flight 1 (Private Pilot Flight 1)	1
AERF 1112	LeTourneau Flight Operations	2
AVTC 2431	Powerplant Electrical Systems Laboratory	1
AVTC 2433	Powerplant Electrical Systems	3
AVTC 2601	Introduction to Aircraft Reciprocating Engines Laboratory	1
AVTC 2603	Introduction to Aircraft Reciprocating Engines	3
BIBL 1043	Biblical Foundations for Living	3
	Semester Total	**14**
Fourth Semester		
AERO 2203	Meteorology for Pilots	3
AERF 1121	Primary Flight 2 (Private Pilot Flight 2)	1
AVTC 2231	Introduction to Aircraft Systems Laboratory	1
AVTC 2232	Introduction to Aircraft Systems	2
AVTC 2441	Airframe Electrical Laboratory	1
AVTC 2443	Airframe Electrical	3
BIBL	Jr.-Sr. Level Elective	3
	General Elective	2
	Semester Total	**16**
Fifth Semester		
AERF 2103	Flight Science II (Instrument Ground School)	3
AERF 2311	Commercial Flight I (Tailwheel Operations)	1
AERO 2112	Aerodynamics for Pilots	2
AERO 3823	Aviation Management	3
ENGL	Literature Elective	3
HIST	History Elective	3

	General Elective	2
	Semester Total	**17**
Sixth Semester		
AERF 3103	Flight Science III (Commercial Ground School)	3
AERF 3412	Instrument Flight	2
AERO 2423	Aviation Human Factors	3
AERO 3213	Electronic Navigation & Flight Control Systems	3
PHYS 1111	General Physics I Laboratory	1
KINE	Physical Activities	1
PHYS 1113	General Physics I	3
	Semester Total	**16**
Seventh Semester		
AERF 2411	Commercial Flight II - Multi-Engine Operations	1
AERO 4133	Crew Resource Management	3
AVTC 3211	Transport Category Aircraft Systems Laboratory	1
AVTC 3212	Transport Category Aircraft Systems	2
AVTC 3612	Introduction to Aircraft Gas Turbines	2
	Lab Science Elective (2 classes)	4
	General Elective	3
	Semester Total	**16**
Eighth Semester		
AERF 3512	Commercial Flight III - Air Transport Emphasis	2
AERO 4123	Commercial Flight Operations	3
AERO 4142	Aviation Law	2
MATH 1503	Calculus with Business Applications	3
BIBL	Jr.-Sr. Level Elective	3
	General Elective	3
	Semester Total	**16**

TAMU-CT
1901 S. Clear Creek Rd.
Killeen, TX 76549
254.519.5469

TAMU-CT is unique in that it offers the only public-assisted, four-year aviation degree program in Texas. It is also the only university in Texas that offers no lower-level classes – only junior, senior, and graduate level courses. All lower-

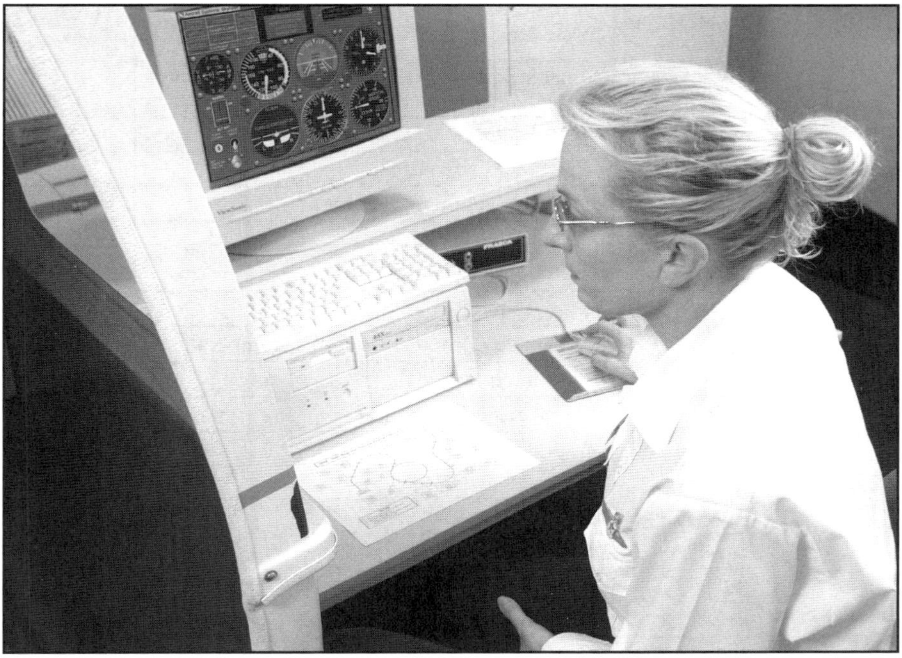

level core and aviation courses, as well as flight training credits, are transferred into TAMU-CT. All freshman and sophomore core curriculum classes – as well as flight courses, ratings, and certificates – must be completed at specified approved colleges outside of TAMU-CT.

	Lower-level courses to be taken at an approved college	
Course		**Credit Hours**
ENGL 111 (1301)	Intro to College Composition	3
ENGL 112 (1302)	College Composition & Research	3
	Communications	3
MATH 107 (1314)	College Algebra or above	3
	Biology, Chemistry, Physics, or Geology	7
	Literature course in English, Sophomore English	3
HIST 201 (1301)	U. S. History through 1877	3
HIST 202 (1302)	U. S. History since Reconstruction	3
POLS 201 (GOVT 2302 or 2305)	American National Government	3

POLS 202 (GOVT 2301 or 2306)	Texas Government	3
	Visual & performing arts course (not including applied or performance)	3
	Social & Behavioral Science	3
	Activity P. E. or Wellness	2
	Core Curriculum Requirements	**42**
	Lower Division Aviation	
Course		**Credit Hours**
CIS 103	Elementary Computer Concepts & Applications	3
AIRP 1255	Flight-Intermediate	3
AIRP 1305	Aircraft Science	3
AIRP 1307	Aviation Meteorology	3
AIRP 1341	Advanced Air Navigation	3
AIRP 1343	Aerodynamics	3
AIRP 1417	Private Pilot Ground School	3
AIRP 1451	Instrument Ground School	3
AIRP 2250	Flight-Instrument	3
AIRP 2337	Commercial Ground School	3
AIRP 2239	Flight-Commercial	3
AVIM 2331	Airline Management	3
	Lower Division Aviation Hours	**36**
	Total Lower Division Hours	**78**

TAMU-CT offers two degree plans: Professional Pilot and Aviation Management.

	Professional Pilot Courses Include	
Course		**Credit Hours**
ASCI 301	Air Carrier Operations	3
ASCI 302	Techniques of Instruction	3
ASCI 303	Air Traffic Control	3
ASCI 304	Airport Management	3
ASCI 401	Aviation Law	3
ASCI 402	Flight Engineer	3
ASCI 408	Aviation Safety	3

	Aviation Management courses taken at TAMU-CT	
Course		**Credit Hours**
CIS 347	Data Communications for Business	3
GB 311	Business Statistics	3
GB 312	Business Correspondence	3
MGMT 407	Business Ethics	3
ADVANCED ELECT	Aviation or Business	10

Baylor University
Baylor Aviation Science
Clifton Robinson Tower
700 S. University Parks Dr.
Waco, TX 76706
254.710.3435

Bachelor of Science in Aviation Sciences

BU does not maintain its own flight school but has agreements with several local flight schools for its students to

complete the required flying courses. A student who wishes to incorporate their flight training fees into their financial aid processing, however, may only complete flight training at TSTC Waco. Choosing any other flight school requires the student to pay the chosen provider without the benefit of financial aid. Students thereby have access to merit-based scholarships, need-based scholarships, grants, part-time student employment, and student loans. As BU is a member of the University Aviation Association, students can also apply for several aviation-based scholarships offered through that organization.

BU Aviation Science students have an opportunity to complete an internship with an established aviation company or firm. Standard internships run during a summer semester of six-twelve weeks or throughout a fall or spring semester of twelve to sixteen weeks. Each internship must be approved by a faculty advisor.

	Thirty-three hours required from TSTC	
Course		**Credit Hours**
AIRP 1255	Intermediate Flight	2
AIRP 1301	Air Navigation	3
AIRP 1417	Private Pilot Ground School	4
AIRP 1451	Instrument Ground School	4
AIRP 2236	Certified Flight Instructor - Airplane	2
AIRP 2239	Commercial Flight	2
AIRP 2251	Multi-Engine Flight	2
AIRP 2250	Instrument Flight	2
AIRP 2331	Advanced Meteorology	3
AIRP 2333	Aircraft Systems	3
AIRP 2337	Commercial Ground School	3
AIRP 2349	Instructor Ground School	3
	Twenty-seven hours required from Baylor	
Course		**Credit Hours**
AVS 1300	Introduction to Aviation	3
AVS 3310	Aviation Law	3
AVS 3311	Aviation Safety	3
AIRP 1307	Aviation Meteorology	3

AVS 4314	Advanced Aircraft Systems	3
AIRP 1343	Aerodynamics	3
	Including nine hours of "3000" or "4000" aviation science courses	
AIRP 2331	Advanced Aviation Meteorology	3
AVS 4309	Progressive Team Organization	3
AVS 4305	Special Topics in Aviation	3
AVS 4314	Advanced Aircraft Systems	3
AVS 4315	Advanced Aerodynamics	3
AVS 4485	Introduction to Geographic Information Systems	4
	Required courses in other fields	
Course		**Credit Hours**
MTH 1321	Calculus I	3
MTH 1322	Calculus II	3
CHE 1302	Basic Principles of Modern Chemistry and Introductory Chemistry Lab	3
PHY 1408	General Physics for Natural and Behavioral Sciences I or General Physics I	4
CSI 3303	Information Technology	3

Additional advanced elective hours should contain content that applies to the aviation industry.

Tuition and Fees

Tuition, housing, and fees are paid up front at the beginning of each semester.

Tuition

Tuition is the money you pay specifically for classes. It varies tremendously between community and state colleges and private universities. It also varies in how it is calculated by the school. Traditionally, tuition is calculated by a cost per semester hour. The cost for an individual class would be the per-hour price times the number of semester hours that class is assigned. Lab classes are usually fewer hours than lecture classes.

Some schools, however, have a flat semester charge for any full-time student class load (usually a minimum of twelve hours and a maximum of eighteen). Part-time students are charged a cost-per-hour rate. With this sort of tuition structure, the more hours you take, the less the cost is for each semester hour.

Tuition at community and state schools is usually lower than at private universities, but the rate depends on a student's residency qualifications. For state universities, rates for out-of-state students are considerably higher than for in-state. Community colleges have the same type of rate system with lower in-district or in-county rates and higher out-of-district or county rates. (A district is considered to be the area in which the college collects taxes for support.) Some community colleges have a three-tier tuition schedule, with the lowest cost for in-district students, higher costs for out-of-district students, and even higher costs for out-of-state students. Residency requirements can be complex and vary by school. For example, at some schools, a student qualifies as a resident if his or her family owns property in the district and pay taxes on it, even if they do not live on that property.

Private universities usually have higher tuition rates than state universities, but the rates typically do not change for out-of-state students.

Tuition rates will often change every year or two, so checking with the school for its current rate, either on its website or with its admissions office, is necessary. Some state schools also may have an additional "designated tuition" charge – a portion of expenses they are allowed to designate as tuition. Even though there may be a set credit hour charges shown in the general information of a website or catalog, make sure you find out the whole anticipated cost for a semester so you can plan accurately for your tuition costs.

Profile: *Ed Viera, APT Instructor*

Born in Puerto Rico, Ed Viera remembers always being fascinated by planes. He took his first flight at ten when his family moved to Brooklyn, New York. The flight only increased his fascination, along with his desire to work in aviation.

At seventeen, Viera joined the Army, where he became a helicopter mechanic. He served fifteen years as a CH-47 (Chinook) helicopter maintenance technician, platform instructor, supervisor, and in-flight crewmember. As an in-flight crewmember, Viera accompanied flights in Korea, Japan, Seattle, and all over the southern United States. If something broke during the flight and required the helicopter to land, Viera was responsible for repairing the aircraft wherever they happened to land. "We've broken down on mountain tops the size of my office and in rice paddies," he says. There were no spare parts carried on the flight, however, so all he had to fix the helicopter with, he says, was "a big toolbox." He went to flight school on the side and earned his pilot's license.

After leaving the service, he worked as a mechanic for a New York City sightseeing helicopter company. All the maintenance and repair work had to be done at night so the aircraft would be ready for the next morning's tours. After two years he moved to Seattle and became a flight instructor and aircraft mechanic.

He continued to earn more advanced flight certifications, and in 1996, moved to Texas to fly freight for a Dallas-Fort Worth cargo company. Acquiring almost 1000 hours of flight time in a turbo-prop Cessna Caravan, he also started and managed a one-man, one-plane independent flight school, training new pilots. After three years, which he describes as "killing myself trying to meet deadlines, flying nights through all kinds of nasty weather, and not having a life," he came to TSTC Waco as a part-time flight instructor in the APT program.

Advancing to CFI in July 2000, Viera now supervises eight flight instructors, coordinates with the FAA, and teaches two aviation meteorology classes. With over 5000 hours of flight time, Viera holds commercial single and multi-engine instrument pilot ratings and a Gold Seal Flight Instructor certificate. He has also earned advanced and instrument ground instructor certificates, an aviation maintenance technician – airframe and powerplant certificate, and associate degrees in Aviation Flight Technology and aviation maintenance. "I just kept going and going," he says.

Viera believes strongly in the value of the training environment he works in. "You can go to your local airport, and they'll be able to teach you to fly," he explains. "But I like the collegiate program – you can't compare the knowledge base and experience you get from the collegiate program."

When it comes to teaching, he says, "I love it. To me this job is just about perfect: the right amount of flying, the right amount of administrative." Working with the flight instructors and students, he performs instructor evaluations and helps students who may be having some trouble acquiring the necessary flight skills.

He cautions, "This is no party school." He has observed that students in their first experience living away from home often see it as a chance to break the rules and abandon discipline. "Those students tend to self-eliminate. They either quit partying or quit the program." He emphasizes, however, that even though some students may be a little slower to develop the skills to be a professional pilot, "the ones that stick with it actually go far."

In some cases, tuition at community and state schools can be waived if a student meets certain requirements. This is decided on a case-by-case basis; check with the schools you research to see if you qualify. There is also a tuition rebate available in some states for qualifying students, but it is paid after finishing, rather than before or while attending school.

Housing

Housing is another element that can affect the cost of attending college. Some schools offer on-campus housing and some do not. Those that do offer on-campus housing sometimes have requirements that first-year students must live on campus. Those students are automatically assessed a dining-hall fee as well as housing charges.

Student Fees

Student fees are the money you pay to colleges and universities for things other than classes. Most schools charge lab fees to cover the cost of materials used in laboratory

classes. Some schools have a list of the individual fees they charge, for example, a student activity fee, a health center fee, a library fee, an ID card fee, a records fee, a parking fee, and computer-access fees. Other schools roll most of the individual fees into an all-inclusive "student fee."

Many state-assisted schools will waive some or all of their fees for students who meet certain qualifications. These can be very complex and detailed. Check school websites and with their financial aid offices to see if you qualify.

Flight Fees

Aviation students have an additional group of fees to cover, which involve the costs of flight hours and flight simulator hours. Unlike regular fees, flight fees are paid on an as-you-go basis. They are a substantial part of the cost of any type of degree that includes achieving professional pilot status, and you should be aware of the cost ahead of time.

Each school will have its own way of handling flight fees. Schools that do not have their own planes may have students pay the flight instructor's school directly. At a college that maintains its own fleet of planes, flight fees will be paid to the college. For example, TSTC Waco has its own fleet, and at the beginning of his or her first semester, each student is required to set up a "flight account" with the Aviation Program. A minimum of $2000 must be deposited into the account at that time. Every time the student practices in a simulator or plane, the charges for that time are deducted from the account. A student must maintain a minimum of $250 in the account to fly, and the account can be refilled at the student accounts office at any time.

APT department chair Bill McNeill recommends that $1200 be deposited in the account every month of classes after the initial deposit. "This allows a student to fly twice a week – eight times a month – which keeps them on course with the curriculum," he says. The account is active until the program is completed, so if the student has not been able to complete the required flying hours at the end of classes, he or she can continue replenishing the account and flying until they achieve their final certifications. According to McNeill, most students do have to continue flying after the scheduled semesters to complete the required hours due to bad weather, planes out of commission, and lack of funds.

The amount deducted from the account for each session depends on the simulator or aircraft used. TSTC Waco has ten planes of four different types and four different simulators for practice. Two hours are blocked out in the schedule for each one-hour flight (or simulation), allowing for preparation and pre-flight checks before an hour of flying time.

Fees for each session include plane rental/fuel use and the flight instructor's time. The most basic training plane runs (in 2010) about $150 per session – $105 for the plane and $45 for the instructor. The other basic plane runs approximately $165 per session. The multi-engine planes run about $185 to $235 per session. (Costs can fluctuate as fuel costs change.) A single-engine Frasca 141 Simulator runs about $70 per hour and a twin-engine Frasca 242 about $95 per hour. (TSTC also has two new Redbird full-motion simulators, which will be used in future training, but a rate schedule had not been established at the time this was written.)

If a student does not show up for a scheduled flight training session, half the charge is deducted from the account as a 'no-show' fee. When a student graduates or withdraws from the APT program, any money left in the account can be returned to the student upon written request. If financial aid is used to help pay flight fees, any remaining money in the student's flight account may be returned to the aid provider(s).

As you can see, flight fees are a very significant part of the cost of either an AAS or bachelor's degree for a professional pilot. At TSTC Waco, flight fees for the complete program run almost $40,000, with tuition, regular fees, and books costing around $9-$10,000. Careful school research will help you adequately prepare for the costs.

Additional Fees (FAA)

In addition to tuition, school fees, and flight fees, there are FAA fees for each flight certification. When a student has completed a flight course and the required flight hours for a specific certification level, he or she must pass a written FAA test and a check flight with an FAA Designated Flight Examiner. There is a $100 fee for the written test and a $300 fee for the check flight. These requirements apply to each new level of certification: private pilot, commercial pilot, instrument rating, and multi-engine rating. There will be a $100 test fee and a $300 check flight fee for each new level.

Other Fee Examples

In addition to tuition, the following fees may be required or applicable. Except for Individualized Instruction, most fees are nonrefundable. Flight fees are refundable based upon actual flight time used.

Return Check Fee	$25
Course Challenge	$50
GED Testing	$60
ACT ASSET Test	$20
FAA Knowledge Test	$55-$125
Instructional Materials (as required by contract, per credit hour) (non-refundable)	$30-$40
Nursing Insurance, Fees and Testing	$20-$99
Student I.D., Meal Card Replacement, or Bus Pass Replacement (per card)	$5
Installment Plan (nonrefundable)	$20
Individualized Instruction Aviation Science	$1,575 - $5,000
Overflight Fees (per hour)	$25 - $165

Two-Year College Fees

For reference purposes, here are examples of cost schedules for several two-year colleges from each institution's website. These costs were current for Fall 2011 (costs vary each year and, at some institutions, each semester).

Central Texas College

Central Texas College Tuition Schedule			
Semester Hours	Resident In-District	Resident Out-of-District	Nonresident International
1	$59	$76	$200
2	$118	$132	$350
3	$177	$228	$525
4	$236	$304	$700
5	$295	$380	$875
6	$354	$456	$1050
7	$413	$532	$1225
8	$472	$608	$1400
9	$531	$684	$1575
10	$590	$760	$1750
11	$649	$836	$1925
12	$708	$912	$2100

(The cost continues increasing at the same rate through 18 semester hours.)

Palo Alto College – San Antonio, TX

The cost of tuition and fees to attend Palo Alto College for each ten-month period is approximately $8,000. Cost related to housing, living expenses, books, and supplies vary from semester to semester and from student to student. (The figures from the table below should be used for approximation only.)

Approximate cost to attend Palo Alto College over a ten-month period:	
Tuition and fees	$8000*
Books/Supplies	$2000*
Housing/Living Expenses	$10,000**
Total	**$20,000**

* These are projected estimates of school tuition, books, and living expenses. Tuition fees and books may vary according to number of registered credit hours.
** Housing and living expenses may vary based upon the location and choice of apartment complex. Additional funds for "Living Expenses" are required if spouse and/or children will accompany the student.

Note that flight fees are not included in the totals shown. This information is not shown on the website and must be obtained from the admissions office.

TSTC Waco

Tuition and flight fees for 2010

Sem	Hrs	Tuition @67.00 Credit Hr	Designated Tuition @34.00 Credit Hr	Textbooks & Supplies	Tools & Equipment	Sem Total
1	15	1,455.00	690.00	520.00	8,000.00	10,665.00
2	15	1,455.00	690.00	370.00	9,200.00	11,715.00
3	14	1,358.00	644.00	220.00	9,200.00	11,422.00
4	14	1,358.00	644.00	190.00	10,255.00	11,859.00
5	14	1,358.00	644.00	255.00	6,300.00	8,557.00
Total	**72**	**6,984.00**	**3,312.00**	**1,555.00**	**41,900.00**	**53,751.00**

Out-of-State Tuition per credit hour: $225.00
Regular fees for students attending TSTC were not listed on the website.

Procedure for Handling Flight Fees

Flight fees can be paid in Student Accounting. An initial deposit of $2,000 starts you flying, with a $250 minimum maintained thereafter. Depositing another $1,200 each month is recommended to allow flying eight times a month and completing the program on time.

Upon graduation or withdrawal from the APT program, unused flight fees can be returned upon written request. Portions of the flight fee deposit from a school grant or loan is cleared through the originating agency to reclaim it. Withdrawals for other purposes are not allowed. Please note that all students enrolling in the APT program must have a second-class medical certificate in their possession at the time of enrollment.

Four-Year College Fees

For reference purposes, here are examples of cost schedules for a couple of four-year colleges from each institution's website. These costs were current for Fall 2011 (costs vary each year and, at some institutions, each semester).

LeTourneau University

Tuition examples

Item	Cost Projection
Tuition/semester (12 - 18 credit hours)	$11,560
Tuition: additional hours beyond 18	$611 per hour
Tuition: 1 - 6 hours	$420 per hour
Tuition: 7 - 11 hours	$922 per hour
Dual Enrollment Per Course	$290 (some limits apply)
Summer School 2010 + Online courses	$342 per hour

Some examples of housing and fees charged:

Item	Fee Projection
Room and Board/semester (with roommate)	$4340
Room and Board/semester (Private Room)	$4840
General Activity Fee (charged to students with 6 or more credit hours)	$110

Campus Post Office Key Deposit	$20
Medical Insurance 1 year United States Resident	$510 (estimated)

Aviation Fees

Aircraft Hourly Rates	**Fee Projection**
Cessna 172 (G-1000)	$138
Citabria (Tailwheel)	$115
Diamond DA-42 (Multi-eng)	$290
Piper PA-32 (Hi-Performance)	$210
Frasca Mentor G-1000 AATD (Simulator)	$48
Frasca DA-42 AATD (Simulator)	$96
Instructor	$57

Fees/Flight fees projected for required courses for Aeronautical Science & Professional Flight degree program:	
Primary Flight 1	$5,965
Primary Flight 2	$7,335
Commercial Flight 1 – Tailwheel Ops	$5,225
Instrument Flight	$7,255

Commercial Flight II – Multi Eng Ops	$12,385
Commercial Flight III - Air Trans	$6,655
Total Required Courses	**$44,820**

Baylor University Undergraduate Student Costs For 2011-2012

Baylor University's costs are the same for in-state students as they are for out-of-state students.

Tuition and Required Fees per year	Amount	Estimate for 12 Hours or more
Tuition, per semester (12 hours or more)	$14,360	$14,360
Less than 12 hours, per semester hour	$1,197	
General Student Fee, 12 hours or more per semester	$1,469	$1,469
General Student Fee, less than 12 semester hours, per semester hour	$123	
Laboratory/Course Fee, (minimum) varies per course	$50.00	$50.00
Chapel Fee	$65	$65
Matriculation Fee (1st time Freshmen and transfer undergraduate students only)	$100	$100
Total required tuition and fees per year		$16,044
Total required tuition and fees for four years		**$82,480**

Because Baylor has a contract with TSTC Waco for aviation student flight training, flight fees of approximately $40,000 (as shown above) from TSTC Waco would be added to this total, bringing the total for the Aviation Sciences degree to approximately $120,000-125,000.

Financial Aid

Financial aid is another area in which community, state, and private colleges and universities have an advantage over private flight schools – more types of aid are available, and there is a financial aid office with employees who can help you find funding sources for your education. Because the possible aid varies by state and school, it is imperative to talk to the financial aid office at the school you are considering.

They will be able to help you determine your eligibility for the various kinds of aid.

There are many types of financial aid available. Federal loans and grants are available nationwide, and many states have additional loan and grant sources. A grant is essentially a gift of money that does not have to be repaid as long as the conditions for receiving the grants are met.

Scholarships also provide money that does not have to be paid back. Many scholarships are, for instance, offered by schools, businesses, industry, and social or professional organizations. For most scholarships, you must meet certain qualifications. Many companies offer one or more scholarships for children of their employees. Professional and social organizations offer scholarships for children of their members. Some scholarships are awarded only to students who are studying in a specific field. Some are reserved for students of a particular national heritage (for example: Polish, Irish, or Hispanic). There are some that aim at helping minorities or women move into technical fields.

There are books with lists of scholarships, which may be available in your local library or book store. Internet websites also allow you to search for scholarships (for example: scholarships.com). It is not necessary to pay a service for an online search; many of them allow you to search for free. Many of the aviation-related websites listed in Chapter Three of this book offer links to sources of aviation scholarships, including the FAA.gov and AOPA.com sites.

Some groups of people qualify for tuition waivers, which are mostly offered at state colleges and universities. There are also possibilities of a tuition rebate in some states, based on certain qualifications. Tuition rebates typically are received at the end of the degree program rather than at the beginning. Financial aid offices at individual schools will have information on which of these may apply to students at their institutions.

Flight Teams

One of the advantages of attending a Part 141 or Part 142 flight school that has its own planes is the possibility of participating in a competitive flight team. The National Intercollegiate Flying Association hosts the flying competition Safety and Flight Evaluation Conference (SAFECON) each year.

Profile: *Samip Shah, APT Student*

Born in India, Samip Shah's first flight was an adventure. Only six or seven years old, he flew unaccompanied from Bombay to New York City. Traveling to meet his family, who had moved to America to start a business, he spent about nineteen hours on a Boeing 747, "an actual double-decker plane," he says. He was fascinated, and by the end of the flight, he says, "I just fell in love with planes. From then on, I knew that I wanted to be a pilot."

Ultimately living in Iowa, Shah attended the University of Iowa for a year, taking pre-med classes because of his parents' concern about the uncertainty of a career in aviation. Unable to resist the pull of flying, he researched the options for getting a professional pilot's certificate and chose a two-year technical school. Major factors in his decision were the relatively short time required to acquire the necessary certifications and ratings, the cost, and the accomplishment of an associate degree.

Shah's biggest surprise when he started the APT program at TSTC Waco was that he started flying right away. "I thought you'd have to take ground courses for like a year first and then start flying, but at TSTC you start off flying. It's great."

No class was particularly hard for him, he says, but in working towards his CFI, he has found it challenging, "Trying to teach people. The hardest thing to teach is aerodynamics because there's so much complex theory you have to learn. There's so much stuff to know about it, and you have to have different ways to teach different people."

Shah expects to complete his flight training in three years rather than the program's scheduled five semesters. "There are times the weather is bad," he explains, "or a plane breaks, or you aren't motivated much." Shah believes motivation is the key to making it through the training. He's seen a number of student pilots drop out because they didn't have it. "It takes a minimum of 250 flight hours to get a commercial certificate. There's a long way to go to get there. A lot of people give up, thinking that it's really hard, but it's really not. There are a lot of things you have to know, but you can learn it."

Being prepared is essential to completing the APT program, Shah emphasizes. "Know how much it's going to cost you. About half of the people who come into flight training aren't prepared with the flight training money. When they find out how much it costs, they drop out. Don't give up," he exhorts. "Have patience. Sometimes you have to wait to fly, and sometimes you have to wait to get a job. And a lot of young pilots have two jobs."

His warning to students considering a career as a professional pilot: "Don't do it for the money. You'll get paid better later on, but that's a sacrifice you have to make at first." As for himself, knowing the challenges he has overcome and the ones still ahead, "I am still happy with my choice. For me it's not about the money."

His own goal is to fly for FedEx someday, preferring cargo to the airline passenger route. "It's not more money," he explains, "but it's more stable." In the end, though, he just wants to fly.

Competitions are fun, challenging, and a great chance to hone both flight and ground skills. Flying to and from the competition, as well as during practice and competition itself, logs hours and adds accomplishments to a resume.

SAFECONs offer both flight and ground competitions. Only student pilots who already have the required certifications for each event can participate in the flying events. For example, students must have an instrument rating to compete in the IFR events. Any student pilot can participate in the ground events, however.

An elimination process is pursued in each school to select its flight team. The teams then compete regionally, and the three highest-place teams in each region move on to the national level. Each event has an individual winner or winners in two-man events. Team standings are determined by the number of events won by team members. The team with the most individual medals wins the team trophy.

Flying events include the following:

- Power Off Landing
- Short Field Approach and Landing
- Navigation
- Message Drop
- IFR Precision Flight Event
- IFR Simulated Flight
- Crew Resource Management/Line Oriented Flight Training

Ground events

- Computer Accuracy
- Ground Trainer
- Aircraft Preflight Inspection
- Aircraft Recognition
- Simulated Comprehensive Aircraft Navigation
- Electronic Flight Computer

Conclusion

Airlines and aviation companies look for the best and brightest for employees, almost requiring that all candidates be highly educated. Many students begin with an associate degree, which allows them to achieve many certifications (commercial single-engine certificate, multi-engine commercial rating, and instrument flight rating, for instance), as well as log hours and acquire experience. A bachelor's degree, however, is becoming more of a necessity in order to compete effectively in the aviation job market, particularly with passenger airlines. A bachelor's degree shows that a student has dedication and a willingness to work hard. It also allows a student to compete on an equal playing field with the other job candidates, especially during times when the economy is tighter and airlines are pickier.

Notes

CHAPTER THREE

Additional Aviation Pilot Information and Resources

"How do you find out if there's a pilot in the crowd? You don't have to – he'll tell you!" – Aviator joke

One of the best resources a pilot has is other pilots. From information regarding flight schools to job openings to aviation companies to new aircraft and technology, pilots invariably love to talk about flying and all things related. Making and maintaining friendships as you work through flight training and professional jobs will be an indispensable tool for your career in aviation.

Flight Schools

There are many flight schools across the country, and they offer many different levels of instruction with varied certifications, ratings, and types of instruction. P61 instruction is an individual course with a looser, less-strict structure. P141 is a structured, classroom-based environment with a detailed syllabus that is reviewed and approved by the Federal Aviation Administration (FAA). P142 is also classroom-based and FAA-approved, but includes newer, more advanced flight simulators with more hours allowed on simulators instead of planes to qualify for certification. There are many different types of simulators available, the most recent and advanced types providing full-motion experience and a "glass cockpit" – screens that mimic the view that would be seen outside a plane's windows and a heads-up LED display is projected onto the cockpit window.

Any of the flight schools can provide a variety of certifications, depending on the skills and certifications of their flight instructors, but the P141 and P142 training environments are more in-depth and require somewhat fewer hours of flight time to complete. Both P141 and 142 schools are heavily supervised and frequently inspected by the FAA on an unannounced basis, checking instruments, books, planes, systems, weather monitoring, and all aviation courses and academic electives. This close supervision by the FAA allows P141 and 142 schools to graduate pilots with fewer flight hours. If you are planning to work towards a commercial certificate and fly professionally ("for hire"), the P141 and P142 courses are often more valued by potential employers and, therefore, may increase your chances of getting the job you want.

This list of schools includes most of the flight schools listed on the FAA website. It also includes additional colleges and universities that offer aviation degree programs, both two- and four-year. You can contact the individual schools or look them up on the Internet to find out what certifications and specialized training they offer. There are more than 4000 flight schools in the United States. Various websites, such as FAA.gov and AOPA.com, offer extensive information and sometimes links to flight schools and allow you to search either by the state or by the name of the school.

Higher Education Aviation Pilot Programs in the United States

ALABAMA

Auburn University
415 W. Magnolia Ave.
Auburn, AL 36830
334.844.4597
flghted@auburn.edu
www.business.auburn.edu/amlg

Double Bridges Aviation
809 Aviation Blvd., Hangar 811
Enterprise, AL 36330
334.393.6774
qjerkins@msn.com
www.doublebridgesaviation.com

Enterprise Ozark Community College
3405 Hwy. 231 S.
Ozark, AL 36360
334.774.5113
www.escc.edu/ozark/tour.htm

Flight Training of Mobile
2241 Michigan Ave.
Mobile, AL 36615
251.441.0723
flighttraining@flyftm.com
www.flyftm.com

Skywarrior, Inc.
12301 Airport Rd.
Bay Minette, AL 36507
251.937.5050
www.skywarriorinc.com

Tuskegee University
1200 W. Montgomery Rd.
Tuskegee, AL 36088
800.622.6531
www.tuskegee.edu

Wallace State Community College
801 Main St. N.W.
Hanceville, AL 35077
256.737.3040
marsha.bradford@wallacestate.edu
www.wallacestate.edu

ALASKA

Aero Tech Flight Service, Inc.
1100 Merrill Field Dr.
Anchorage, AK 99501
907.279.6558
info@aerotechalaska.com
www.aerotechalaska.com

Elmendorf Aero Club
P.O. Box 6292
JBER, AK 99506
907.753.4167
jhfred@mtaonline.net
www.elmendorfaeroclub.com

Land & Sea Aviation Alaska, LLC
2400 E. 5th Ave.
Anchorage, AK 99501
907.274.2544
lasak@clearwire.net
www.landandseaaviationalaska.com

Take Flight Alaska
1740 E. Fifth Ave.
Anchorage, AK 99501
907.274.9943
tfa@takeflightalaska.com
www.takeflightalaska.com

University of Alaska Anchorage
2811 Merrill Field Dr.
Anchorage, AK 99501
907.786.7200
ayflyhi@uaa.alaska.edu
www.uaa.alaska.edu/aviation

Warbelow's Air Ventures
3758 University Ave. S.
Fairbanks, AK 99709
907.474.0518
david.spence@warbelows.com
www.warbelows.com

ARIZONA

APS Emergency Maneuver Training
5865 S. Sossaman Rd.
Mesa, AZ 85212
866.539.4273
info@apstraining.com
www.apstraining.com

Arizona Aero-Tech
6971 S. Apron Dr.
Tucson, AZ 85756
520.294.1551
www.arizonaaerotech.com

Arizona State University
7442 E. Tillman Ave.
Mesa, AZ 85212
480.727.1024
aerotech@asu.edu
www.technology.asu.edu/aviation

Chandler Air Service, Inc.
1675 E. Ryan Rd.
Chandler, AZ 85286
480.963.6420
fly@aerobatics.com
www.aerobatics.com

Chandler-Gilbert Community College
2626 E. Pecos Rd.
Phoenix, AZ 85225
480.732.7000
cindy.clark@cgcmail.maricopa.edu
www.cgc.maricopa.edu/aviation

Cochise Community College
4190 W. Hwy. 80
Douglas, AZ 85607
520.364.7943
aviation@cochise.edu
www.cochise.edu/deptsdirs/aviation/index.asp

Embry-Riddle Aeronautical University
3700 Willow Creek Rd.
Prescott, AZ 86301
928.777.6600
pradmfrm@erau.edu
www.erau.edu/pr

Falcon Executive Aviation
4766 E. Falcon Dr.
Mesa, AZ 85215
480.832.0704
www.falconaviation.com

K&S Aviation Services
4835 E. Indigo St., Ste. 103
Mesa, AZ 85205
480.398.1260
www.kandsaviation.com

Mesa Pilot Development
7442 E. Tillman Ave.
Mesa, AZ 85212
480.727.1894
tech.mesa-air.com/mpd

Oxford Aviation Academy
1658 S. Litchfield Rd.
Goodyear, AZ 85338
623.925.2090
enquiries@oaa.com
www.oaa.com

Sabena Flight Academy
5010 E. Falcon Dr.
Mesa, AZ 85215
480.948.4515
www.sabena-az.com

Sonoran Wings Flight Training Centre
6720 S. Plumer
Tucson, AZ 85756
520.807.5900
info@sonoran-wings.com
www.sonoran-wings.com

Tiffin Aviation Services Inc.

11 Access Way
Nogales, AZ 85621
520.287.9120
info@tiffinaviation.com
www.tiffinaviation.com

Transpac Aviation Academy

530 W. Deer Valley Rd., Ste. 100
Phoenix, AZ 85027
623.580.7900
info@transpacacademy.com
www.transpacacademy.com

UND Aerospace Flight Training Center

5733 S. Sossaman Rd.
Mesa, AZ 85212
480.988.8117
www.undaerospace.com

Westwind School of Aeronautics

732 W. Deer Valley Rd.
Phoenix, AZ 85027
623.869.6973
www.westwindaviation.com

ARKANSAS

Central Flying Service, Inc.

1501 Bond St.
Little Rock, AR 72202
501.375.3245
www.central.aero

Henderson State University
1100 Henderson St.
Arkadelphia, AR 71999
870.230.5000
aviation@hsu.edu
www.hsu.edu/aviation

Skyventure Aviation, Inc.
3901 Lindbergh Dr.
Jonesboro, AR 72401
870.268.1900
www.skyventureinc.com

Summit Aviation, LLC
2500 S.W. Aviation St.
Bentonville, AR 72712
479.254.0817
davepowell@IFLYSummit.com
www.summitaviationllc.com

CALIFORNIA

Aero Tech Academy, Inc.
1745 Sessums Dr.
Redlands, CA 92374
909.794.4046
info@ata.bz
ata.bz

Ahart Aviation Services
186 Airway Blvd.
Livermore, CA 94551
925.449.2142
ahart@ahart.com
www.ahart.com

American Academy of Aeronautics
Carl Mather Field
Mather, CA 95655
800.437.0527
info@americanacademy.net
www.americanacademy.net

American Aviation Academy, Inc.
2035 N. Marshall Ave.
El Cajon, CA 92020
619.448.9149
info@flyaaa.com
www.flyaaa.com

American Flyers Flight School
2501 Airport Ave.
Santa Monica, CA 90405
800.233.0808
wes@af.tv
www.americanflyerssantamonica.net

Aviation Facilities
4119 W. Commonwealth Ave.
Fullerton, CA 92833
714.773.0741
learntofly@flyafi.com
www.flyafi.com

Aviation Pacific, Inc.
575 Aviation Dr.
Camarillo, CA 93010
805.482.8100
fly@aviationpacific.com
www.aviationpacific.com

Bel-Air International
795 Sky Way
San Carlos, CA 94070
650.596.9900
info@belairintl.com
www.belairintl.com

California Airways, Inc.
22693 Hesperian Blvd., Ste. 220
Hayward, CA 94541
510.887.7686
www.california-airways.com

California Flight Academy, LLC
2065 N. Marshall Ave.
El Cajon, CA 92020
619.448.2212
info@flycfa.com
www.flycfa.com

California State University
5151 State University Dr.
Los Angeles, CA 90032
323.343.4568
kmew@calstatela.edu
www.calstatela.edu

Channel Islands Aviation
305 Durley Ave.
Camarillo, CA 93010
805.987.1301
cia@flycia.com
www.flycia.com

Coast Flight Training and Management Inc.

3753 John J. Montgomery Dr.
San Diego, CA 92123
858.279.4359
www.iflycoast.com

Cypress College

9200 Valley View St.
Cypress, CA 90630
714.484.7411
kreiland@cypresscollege.edu
www.cypresscollege.edu/~aviation

Edwards Aero Club Flight Training Center

5 Seller Ave., Bldg. 3000
Edwards AFB, CA 93524
661.275.2376
www.edwardsaeroclub.org

Executive Flyers Inc.

6151 Freeport Blvd., Ste. 151
Sacramento, CA 95822
916.427.1888
www.execflyers.com

Glendale Community College

1500 N. Verdugo Rd.
Glendale, CA 91208
818.240.1000
cpotter@glendale.edu
www.glendale.edu

Golden State Flying Club
1640 N. Johnson Ave.
El Cajon, CA 92020
619.449.0611
www.gsfc.com

Grey Eagle Flight Academy
2186 Palomar Airport Rd.
Carlsbad, CA 92011
760.804.8670
greyeagleaviation@sbcglobal.net
www.greyeagleflightacademy.com

Group 3 Aviation, Inc.
16425 Hart St.
Van Nuys, CA 91406
818.994.9376
group3@group3aviation.com
www.group3aviation.com

IASCO Flight Crew Training Division
2000 Airport Rd.
Napa, CA 94558
707.252.3522
tng@iaso.com
www.iasco.com

International Flight Training Academy
1450 Boughton Dr.
Bakersfield, CA 93308
661.391.1100
www.ifta.aero

JC Air Academy Inc.
6364 S. Lindbergh St.
Stockton, CA 95206
209.983.2773
www.jcairacademy.com

Jet Aeronautical, LLC
1950 Cordell Court, No.103
El Cajon, CA 92020
619.562.1950
www.jetaeronautical.com

Long Beach Flight Academy
2631 E. Spring St.
Long Beach, CA 90806
562.290.0321
Lbflying@yahoo.com
www.lbflying.com

Mazzei Flying Service
4885 E. Shields Ave., Ste. 201
Fresno, CA 93726
559.251.7501
learn2fly@flymfs.com
www.flymfs.com

Mendocino College
1000 Hensley Creek Rd.
Ukiah, CA 95482
707.468.3000
www.mendocino.edu

Miramar College
10440 Black Mountain Rd.
San Diego, CA 92126
619.388.7664
www.sdmiramar.edu

Mount San Antonio College
1100 N. Grand Ave.
Walnut, CA 91789
909.594.5611
rrogus@mtsac.edu
aeronautics.mtsac.edu

National Air College
3760 Glenn Curtis Dr.
San Diego, CA 92123
858.279.4595
nac@nethere.com
www.nationalaircollege.com

Nice Air Aviation
2575 Robert Fowler Way
San Jose, CA 95148
408.729.3383
info.niceairaviation@gmail.com
www.niceairaviation.com

Orange Coast College
2701 Fairview Rd.
Costa Mesa, CA 92626
714.432.5072
ryoung@occ.cccd.edu
www.orangecoastcollege.edu

Orange County Flight Center
19531 Campus Dr., Ste. 4
Santa Ana, CA 92707
949.756.1300
info@ocflightcenter.com
www.ocflightcenter.com

Pacific States Aviation
51 John Glenn Dr.
Concord, CA 94520
925.685.4400
www.pacificstatesaviation.com

Pacific Union College
One Angwin Ave.
Angwin, CA 94508
707.965.6219
www.puc.edu/angwinairport

Palomar Community College
1140 W. Mission Rd.
San Marcos, CA 92069
760.744.1150
www.palomar.edu/aviation

Pinnacle Aviation Academy
2210 Palomar Airport Rd.
Carlsbad, CA 92011
760.929.1009
www.pinnacleacademy.com

Rainbow Air Academy
2601 E. Spring St.
Long Beach, CA 90806
562.427.8282
www.rainbowair.com

Sacramento City College
3835 Freeport Blvd.
Sacramento, CA 95822
916.558.2278
www.scc.losrios.edu

San Bernardino Valley College
701 S. Mount Vernon Ave.
San Bernardino, CA 92410
909.384.4400
www.valleycollege.edu

San Diego Christian College
2100 Greenfield Dr.
El Cajon, CA 92019
619.201.8700
www.sdcc.edu

San Diego Flight Training International
8745 Aero Dr., Ste. 103
San Diego, CA 92123
858.569.1822
info@sdfti.com
www.sdfti.com

Santa Rosa Junior College
150 Mendocino Ave.
Santa Rosa, CA 95401
707.527.4011
www.santarosa.edu

Scandinavian Aviation Academy
1835 N. Marshall Ave.
El Cajon, CA 92020
619.631.0323
www.bfsaa.se

Sky Walk Flight Training Academy
6151 Freeport Blvd., Ste. 158
Sacramento, CA 95822
916.391.1957
information@skywalkinc.com
www.skywalkinc.com

Squadron-2
2655 Robert Fowler Way
San Jose, CA 95148
408.272.0518
info@squadron2.com
www.squadron2.com

Sunrise Aviation
19531 Campus Dr., Ste. 7
Santa Ana, CA 92707
800.717.4200
info@sunriseaviation.com
www.sunriseaviation.com

The Flight School at Palm Springs, Inc.
210 N. El Cielo Rd.
Palm Springs, CA 92262
760.416.2188
inquiry@flypalmsprings.com
www.flypalmsprings.com

Trade Winds Aviation
2505 Cunningham Ave.
San Jose, CA 95148
408.729.5100
mail@tradewindsaviation.com
www.tradewindsaviation.com

TwinAir Inc.
265 Durley Ave., Ste. 108-113
Camarillo, CA 93010
818.988.7573
www.twinair.net

Universal Air Academy
4233 Santa Anita Ave.
El Monte, CA 91731
626.454.5254
info@universalairacademy.com
www.universalairacademy.com

COLORADO

A-Cent Aviation
1945 Aviation Way
Colorado Springs, CO 80916
719.573.2236
www.a-centaviation.com

Aims Community College
5401 W. 20th St.
Greeley, CO 80634
970.330.8008
aviation@aims.edu
www.aims.edu

Colorado Northwestern Community College
500 Kennedy Dr.
Rangely, CO 81648
800.562.1105
jason.herkimer@cncc.edu
www.cncc.edu

Flights, Inc.
7355 S. Peoria St.
Englewood, CO 80112
303.799.5140
info@flights-inc.com
www.flights-inc.com

Journeys Aviation, Inc.
11855 Airport Way
Broomfield, CO 80021
303.466.8566
www.journeysaviation.com

McAir Aviation, LLC
11945 Airport Way
Broomfield, CO 80021
303.466.8730
www.mcairaviation.com

Metropolitan State College of Denver
1201 5th St.
Denver, CO 80217
303.556.2983
www.mscd.edu/~aviation

Rocky Mountain USAF Flight Training Center
Hangar 133
Peterson AFB, CO 80914
719.556.4310
www.rockymountainflight.com

USAF Academy Aero Club
9222 Airfield Dr.
USAF Academy, CO 80840
719.333.4542
www.usafaservices.com/hobbies-skills/aero-club

Western Air Flight Academy
11915 Airport Way, Ste. 4
Broomfield, CO 80021
303.466.6998
www.flywafa.com

CONNECTICUT

Connecticut Flight Academy
20 Lindbergh Dr.
Hartford, CT 06114
860.722.9667
info@ctflightacademy.com
www.ctflightacademy.com

Interstate Aviation Inc.
62 Johnson Ave.
Plainville, CT 06062
860.747.5519
flyiailynn@aol.com
www.interstateaviation.com

Naugatuck Valley Community College
750 Chase Pkwy.
Waterbury, CT 06708
203.575.8040
bsimon@nvcc.commnet.edu
www.nvcc.commnet.edu

Premier Flight Center, LLC
58 Lindbergh Dr.
Hartford, CT 06114
860.724.2245
info@premierflightct.com
www.premierflightct.com

DELAWARE

Delaware State University
1200 N. DuPont Hwy.
Dover, DE 19901
302.857.6710
www.desu.edu/airway-science-dept

FLORIDA

424 Aviation
13230 S.W. 132 Ave., Ste. 20
Miami, FL 33186
786.242.4848
fly@424aviation.com
www.424aviation.com

ADF Airways
14532 S.W. 129 St., Bldg. 227
Miami, FL 33186
305.233.6648
www.adfairways.net

Aerosim Flight Academy
2700 Flight Line Ave.
Sanford, FL 32773
407.330.7020
www.aerosim.com/academy

Air Orlando Flight School
319 N. Crystal Lake Dr.
Orlando, FL 32803
407.896.0721
www.flyairorlando.com

Airborne Systems Flight Training
2011 S. Perimeter Rd., Ste. G & H
Fort Lauderdale, FL 33309
954.776.0543
www.airbornesystems.net

American Flyers Inc.
801 N.E. 10th St.
Pompano Beach, FL 33060
954.785.1450
BenP@af.tv
www.americanflyers.net

AMS Aviation, Inc.
5600 N. Airport Rd.
Milton, FL 32583
850.623.4151
www.flymilton.com

Angley College
1700 S. Woodland Blvd.
Deland, FL 32720
877.483.3635
www.angley.edu

Avia Flight Academy International Inc.
281 Old Moody Blvd.
Palm Coast, FL 32164
386.586.3380
www.iflyavia.com

Aviator College of Aeronautical Science & Technology
3800 Saint Lucie Blvd.
Fort Pierce, FL 34946
772.466.4822
www.flyaviator.com

Bristow Academy Inc.
365 Golden Knights Blvd.
Titusville, FL 32780
321.385.2919
www.bristowgroup.com/about/academy.php

Broward College Aviation Institute
7200 Pines Blvd., Bldg. 99
Pembroke Pines, FL 33024
954.201.8087
www.broward.edu/aviation

Cams Flight
14421 Airport Pkwy.
Clearwater, FL 33762
727.507.8881
info@camsflight.com
www.camsflight.com

Cirrus Aviation Flight Training
8191 N. Tamiami Trail, No. 107
Sarasota, FL 34243
941.360.9074
www.cirrusaviation.com

Cloud Dancer Aviation
1400 Flightline Blvd., Ste. B
Deland, FL 32724
386.238.7270
www.flycda.com

Crystal Aero Group, Inc.
718 N. Lindbergh Dr.
Crystal River, FL 34429
352.795.6868
www.crystalaerogroup.com

Dean International Inc.
14150 S.W. 129th St.
Miami, FL 33186
305.259.5611
rdean@flymiami.com
www.flymiami.com

Eglin Aero Club
200 Escambia
Eglin AFB, FL 32542
850.882.5148
www.eglinaeroclub.com

Embry-Riddle Aeronautical University
600 S. Clyde Morris Blvd.
Daytona Beach, FL 32114
386.226.6100
www.erau.edu

Endeavour Flight Training, Inc.
14970 N.W. 42nd Ave., Ste. 117
Opa-Locka, FL 33054
305.769.2779
www.e-flighttraining.com

Epic Aviation
600 Skyline Dr.
New Smyrna Beach, FL 32168
386.409.5583
www.epicaviation.com

Europe-American Aviation
200 Aviation Dr. N., Ste. 6
Naples, FL 34104
239.430.9220
www.skystead.com

FIT Aviation, LLC
801 Harry Goode Way
Melbourne, FL 32901
321.674.6500
www.fitaviation.com

FlightSafety Academy
2805 Airport Dr.
Vero Beach, FL 32960
772.564.7600
academy@flightsafety.com
www.flightsafetyacademy.com

Flight Training Professionals
391 Herndon Ave., Ste. A
Orlando, FL 32803
407.896.0077
www.ftpros.com

Florida Aviation Academy
1401 N.E. Tenth St.
Pompano Beach, FL 33060
954.788.3887
www.flaviationacademy.com

Florida Aviation Career Training, Inc.
4900 U.S. 1 N., Ste. 200
St. Augustine, FL 32095
904.824.9401
www.florida-aviation.com

Florida Flight Training Center, Inc.
150 E. Airport Rd.
Venice, FL 34285
941.484.3771
www.fftc.info

Florida Institute of Technology
150 W. University Blvd.
Melbourne, FL 32901
321.674.7619
coa.fit.edu

Florida Memorial University
15800 N.W. 42nd Ave.
Miami Gardens, FL 33054
305.623.3600
www.fmuniv.edu

Florida State College at Jacksonville
5640 New World Ave.
Jacksonville, FL 32221
904.779.4200
rgrace@fscj.edu
www.fscj.edu

Jacksonville University
2500 University Blvd. N.
Jacksonville, FL 32221
800.225.2027
www.ju.edu

Jax Navy Flying Club
P.O. Box 127
NAS Jacksonville, FL 32212
904.777.8549
www.jaxnfc.net

London Aviation Group
100 Aviation Dr. S.
Naples, FL 34104
239.643.4468
www.londonaviation.com

Lynn University
3601 N. Military Trail
Boca Raton, FL 33431
561.237.7000
www.lynn.edu

Miami Dade College
11011 S.W. 104th St.
Miami, FL 33196
305.237.8888
www.mdc.edu/homestead/aviation

Miracle Strip Aviation
1001 Airport Rd.
Destin, FL 32541
850.837.6135
www.miraclestripaviation.com

Naples Air Center, Inc.
230 Aviation Dr. S.
Naples, FL 34104
239.643.1717
www.naples-air-center.com

North Florida Flight Training, Inc.
855-1 St. John's Bluff Rd.
Jacksonville, FL 32225
904.642.3912
www.flynorthflorida.com

NS Aviation
601 S.W. 77th Way, Ste. E
Pembroke Pines, FL 33023
954.964.1276
www.nsaviation.com

Orlando Flight Training
301 N. Dyer Blvd., Ste. 102
Kissimmee, FL 34741
407.518.7766
www.flyoft.com

Ormond Beach Aviation/Euro American School of Aviation
770 Airport Rd.
Ormond Beach, FL 32174
386.673.9862
www.flyoba.com

Palm Beach Flight Training
2633 Lantana Rd., No. 42
Lantana, FL 33462
561.963.8821
www.pbflight.com

Palm Beach State College
4200 Congress Ave.
Lake Worth, FL 33461
561.868.3350
www.palmbeachstate.edu

Paragon Flight Training
511 Danley Dr.
Fort Myers, FL 33907
239.274.3170
www.paragonflight.com

Paris Air, Inc.
3300 Airport W. Dr.
Vero Beach, FL 32960
772.770.2708
www.parisair.com

Pelican Flight Training Center, Inc.
1601 S.W. 75th Ave.
Pembroke Pines, FL 33023
954.966.9750
www.pelican-airways.com

Pensacola Aviation Center
4145 Jerry L. Maygarden Rd.
Pensacola, FL 32504
850.434.0636
www.pensacolaaviation.com

Phoenix East Aviation, Inc.
561 Pearl Harbor Dr.
Daytona Beach, FL 32114
800.868.4359
www.pea.com

Professional Flight Training
1815 N.W. 51st Pl.
Ft. Lauderdale, FL 33309
877.938.3043
info@pft-inc.com
www.pft-inc.com

Silver Express Company
14569 S.W. 127th St.
Miami, FL 33186
305.255.8753
www.silverexpress.com

Skywarrior, Inc.
4345 Jerry L. Maygarden Rd.
Pensacola, FL 32504
850.433.6115
www.skywarriorinc.com

Sterling Flight Training
855-21 St. John's Bluff Rd.
Jacksonville, FL 32225
904.642.9683
www.sterlingflight.com

Sunrise Aviation
740 Airport Rd.
Ormond Beach, FL 32174
386.677.5724
www.flysunriseaviation.com

Tailwheels Etc.
3000 21st St. N.W.
Winter Haven, FL 33881
863.327.6880
www.tailwheelsetc.com

University Air Center
4701 N.E. 40th Terrace
Gainesville, FL 32609
352.335.4681
www.universityaircenter.com

GEORGIA

American Air Flight Training
2000 Airport Rd., Ste. 109
Atlanta, GA 30341
770.455.4203
www.fly-aaft.com

American Flyers
1950 Airport Rd.
Atlanta, GA 30341
866.346.0808
www.americanflyers.net

CSG Aviation

3250 W. Britt David
Hangar 12 Rd.
Columbus, GA 31909
706.324.2453
grussell@csgaviation.com
www.csgaviation.com

Falcon Aviation Academy

95 E. Aviation Way
Newnan, GA 30263
770.486.5561
www.falconaviationacademy.com

Fulton Aviation, LLC

3984 Aviation Cr., Ste. 200
Atlanta, GA 30336
678.472.4365
brant@fultonaviation.com
www.fultonaviation.com

Pegasus Air, LLC

109 Airport Dr.
Statesboro, GA 30461
912.764.2836
pegair@frontiernet.net
www.pegasus-air-rental.com

Quality Aviation, Inc.

PDK Airport
1951 Airport Dr., Ste. 100
Atlanta, GA 30341
770.457.6215
fly@qualityaviation.net
www.qualityaviation.net

The Flight School of Gwinnett
800 Airport Rd., Ste. 100
Lawrenceville, GA 30046
770.513.0000
kim@theflightschool.com
www.theflightschool.com

HAWAII

Hawaii Flight Academy
73-300 U'u St.
Kailua-Kona, HI 96740
808.329.0018
fly@hawaiiflightacademy.com
www.hawaiiflightacademy.com

Moore Air, Inc.
90 Nakolo Pl., Ste. 24
Honolulu, HI 96819
808.833.5628
info@mooreair.com
www.mooreair.com

IDAHO

Jetstream Aviation Inc.
3653 Rickenbacker St.
Boise, ID 83705
208.345.3730
fly@jetstreamaviation.com
www.jetstreamaviation.com

ILLINOIS

American Flyers Inc.
3N040 Powis Rd.
West Chicago, IL 60185
800.323.0808
www.americanflyerschicago.net

Great River Aviation, LLC
1647 Hwy. 104
Quincy, IL 62305
217.885.3353
www.curtisaviationservices.com

Kishwaukee College
21193 Malta Rd.
Malta, IL 60150
815.825.2086
www.kishwaukeecollege.edu

Lewis University
One University Pkwy.
Romeoville, IL 60446
800.897.9000
www.lewisu.edu

McClelland Aviation Company, Inc.
1100 Aviation Ln.
Springfield, IL 62707
217.544.9027
www.mcclellandaviation.com

Quincy University
1800 College Ave.
Quincy, IL 62301
212.222.8020
www.quincy.edu

Recurrent Training Center, Inc.
315 S. Dunlap
Savoy, IL 61874
800.727.1014
train@rtcpilot.com
www.rtcpilot.com

Rotors & Wings Aviation, LLC
809 E. 4000 S. Rd.
Kankakee, IL 60901
815.939.3575
tim@rotorsandwings.us
www.rotorsandwings.us

Southern Illinois University – Carbondale
665 N. Airport Rd.
MC 6817
Carbondale, IL 62901
618.453.1147
www.aviation.siu.edu

Southwestern Illinois College
2500 Carlyle Ave.
Belleville, IL 62221
618.235.2700
keith.mueller@swic.edu
www.southwestern.cc.il.us/APT

University of Illinois Institute of Aviation
1 Airport Rd.
Savoy, IL 61874
800.527.8429
www.aviation.uiuc.edu

INDIANA

Eagle Aircraft
4001 Murvihill Rd.
Valparaiso, IN 46383
219.464.0132
oms@jseagle.com
www.jseagle.com

Griffith-Merrillville Aviation
1705 E. Main St.
Griffith, IN 46319
219.924.0207
griffithflightschool@hotmail.com
www.griffithairport.com

Indiana State University
650 Cherry St.
Terre Haute, IN 47809
812.237.2641
patty.guess@indstate.edu
www.indstate.edu/aviation

New Horizons Aviation
17229 Cr. 42
Goshen, IN 46526
574.971.4221
nha@flynewhorizons.com
www.flynewhorizons.com

Vincennes University
Aviation Technology Center
2175 S. Hoffman Rd.
Indianapolis, IN 46241
317.381.6000
dbandy@vinu.edu
www.aviationtechcenter.com

IOWA

Advanced Air, Inc.
16801 McCandless Ln.
Council Bluffs, IA 51503
712.323.2173
advancedaircb@yahoo.com
www.advancedaircb.com

Carver Aero, Inc.
9230 N. Harrison St.
Davenport, IA 52806
563.391.5650
www.carveraero.com

Indian Hills Community College
14802 Terminal Ave.
Ottumwa, IA 52501
641.683.4254
www.indianhills.edu

Storm Flying Service
1524 240th St.
Webster City, IA 50595
515.832.3723
www.stormflyingservice.com

University of Dubuque
2000 University Ave.
Dubuque, IA 52001
563.589.3000
www.dbq.edu

KANSAS

BJS Pilot Ground School
One New Century Pkwy., Ste. 113
New Century, KS 66031
913.780.6444
www.bjspilotgroundschool.com

Central Christian College of Kansas
1200 S. Main
McPherson, KS 67460
800.835.0078
www.centralchristian.edu

Cessna Employees Flying Club
1780 Airport Rd.
Wichita, KS 67209
316.633.4766
www.cessnaflyingclub.org

Hesston College Aviation
325 S. College Dr.
Hesston, KS 67062
316.283.6161
www.hesston.edu

Kansas State University – Salina
2310 Centennial Rd.
Salina, KS 67401
785.826.2679
www.salina.k-state.edu

KENTUCKY

Aero-Tech, Inc.
4330 Hangar Dr.
Lexington, KY 40510
859.254.8906
www.aerotech.net

Air Center 1
2815 Taylorsville Rd.
Louisville, KY 40205
502.451.6451
www.aircenter1.com

Eastern Kentucky University
521 Lancaster Ave.
Richmond, KY 40475
859.622.1014
www.eku.edu

LOUISIANA

Flight Academy of New Orleans, LLC
6101 Giuseppe Bellanca St., Ste. 200
New Orleans, LA 70126
877.359.6652
info@877flynola.com
www.877flynola.com

Louisiana Tech University
Hale Hall
Ruston, LA 71272
318.257.3036
bulldog@latech.edu
www.latech.edu/aviation

University Of Louisiana at Monroe
700 University Ave.
Monroe, LA 71209
318.342.1000
cba.ulm.edu/aviation.html

MAINE

Maine Instrument Flight, Inc.
215 Winthrop St.
Augusta, ME 04330
207.622.1211
www.mif.aero

MARYLAND

Brett Aviation, Inc.
701 Wilson Point Rd.
Middle River, MD 21220
410.391.0210
www.brettaviation.com

Community College of Baltimore County
800 S. Rolling Rd.
Catonsville, MD 21228
410.455.4157
dwilliams@ccbcmd.edu
www.ccbcmd.edu/aviation

Frederick Flight Center, Inc.
330 Aviation Way
Frederick, MD 21701
800.355.0620
lrichardson@frederickflightcenter.com
www.frederickflightcenter.com

Freeway Airport Inc.
3900 Church Rd.
Mitchellville, MD 20721
301.390.6424
flying@freewayaviation.com
www.freewayaviation.com

Navy Annapolis Flight Center
3090 Solomons Island Rd.
Hangar D
Edgewater, MD 21037
410.956.8751
nafc_hq@nafcflying.org
www.nafcflying.org

Trident Aircraft
Bay Bridge Airport (W29)
206 Airport Rd.
Stevensville, MD 21666
410.604.1333
FLY@TridentAircraft.com
www.TridentAircraft.com

University of Maryland Eastern Shore
30806 University Blvd. S.
Princess Anne, MD 21853
410.651.6489
aviation@umes.edu
www.umes.edu

MASSACHUSETTS

Bridgewater State College
95 Grove St.
Bridgewater, MA 02325
508.531.1779
L1Shaw@bridgew.edu
www.bridgew.edu/aviation

East Coast Aero Club, Inc.
200 Hanscom Dr., Ste. 111
Bedford, MA 01730
781.274.6322
ecacbed@eastcoastaeroclub.com
www.ecacbed.com

Executive Flyers Aviation
200 Hanscom Dr.
Bedford, MA 01730
781.274.7227
www.executiveflyers.com

Hanscom Flight Training Center
7 Robbins Rd.
Hanscom AFB, MA 01731
781.377.5160
hanscom.services.online@hanscom.af.mil
www.hanscomservices.com/AeroClub.asp

North Shore Community College
1 Ferncroft Rd.
Danvers, MA 01923
978.762.4000
jbosco@northshore.edu
www.northshore.edu

Northampton Airport
160 Old Ferry Rd.
Northampton, MA 01060
413.584.7980
www.northamptonairport.com

Westfield Flight Academy
111 Airport Rd.
Westfield, MA 01085
413.568.5800
info@fivestarflight.com
www.fivestarflight.com

MICHIGAN

Andrews University Air Park
3898 Griggs Rd.
Berrien Springs, MI 49104
269.471.3547
airinfo@andrews.edu
www.andrews.edu/cot/aviation

Blue Sun Air LLC
5923 Byron Rd.
Zeeland, MI 49464
616.772.5970
amanda@bluesunair.com
www.bluesunair.com

Delta College
1961 Delta Rd.
University Center, MI 48710
989.686.9245
www.delta.edu

Executive Air Transport, Inc.
103 Sinclair Dr.
Muskegon, MI 49441
231.798.2126
www.ea-transport.com

Jackson Community College
3610 Wildwood
Jackson, MI 49202
517.787.7012
www.jccmi.edu

Lansing Community College
419 N. Washington Sq.
Lansing, MI 48901
517.483.1343
www.lcc.edu

Northwestern Michigan College
2600 Aero Park Dr.
Traverse City, MI 49686
231.995.1220
www.nmc.edu/aviation

School of Missionary Aviation Technology
468 E. Sprague Rd. D-175
Ionia, MI 48846
616.527.4160
info@smat-aviation.org
www.smat-aviation.org

Western Michigan University College of Aviation
237 N. Helmer Rd.
Battle Creek, MI 49037
269.964.6375
www.wmich.edu/aviation

MINNESOTA

Academy College
1101 E. 78th St., Ste. 100
Minneapolis, MN 55420
952.851.0066
www.academycollege.edu

Lake Superior College
2101 Trinity Rd.
Duluth, MN 55811
800.432.2884
www.lsc.edu

Minnesota State University – Mankato
328 Armstrong Hall
Mankato, MN 56001
507.389.6116
www.mnsu.edu/programs/aviation.html

MN Aviation, Inc.
400 Airport Rd.
Albert Lea, MN 56007
888.829.5216
www.mnaviation.info

Thunderbird Aviation
5800 Crystal Airport Rd.
Crystal, MN 55429
763.533.4162
www.thunderbirdaviation.com

Twin Cities Flight Training
270 Airport Rd.
St. Paul, MN 55107
651.290.2055
www.tcaviation.com

University of Minnesota – Crookston
2900 University Ave.
Crookston, MN 56716
800.862.6466
www.umcrookston.edu/aviation

MISSISSIPPI

Air Venture Flight Center
11299 Airport Rd.
Olive Branch, MS 38654
901.521.1068
khammerton@aol.com
www.air-venture.com

Delta State University Commercial Aviation
Hwy. 8 W.
Cleveland, MS 38733
662.846.4205
tsledge@dsu.deltast.edu
www.deltastate.edu/pages/457.asp

MISSOURI

ATD Flight Systems, LLC
601 Lou Holland Dr.
Kansas City, MO 64116
816.221.8455
info@atdflightsystems.com
www.atdflightsystems.com

Career Pilot School
104 S. Forrest Ave.
Liberty, MO 64068
877.359.4640
www.careerpilotschool.com

Mizzou Aviation
5497 Dennis Weaver Dr.
Joplin, MO 64801
417.623.1331
www.mizzouaviation.com

Parks College of Engineering Aviation & Tech
3450 Lindell Blvd.
St. Louis, MO 63103
314.977.8203
parks.slu.edu

St. Charles Flying Service, Inc.
6016 Portage Rd.
Portage des Sioux, MO 63373
636.946.6066
www.stcharlesflyingservice.com

University of Central Missouri Flight Facility
TR Gaines 210
Warrensburg, MO 64093
660.543.4975
www.ucmo.edu

MONTANA

Northern Skies Aviation, Inc.
3900 Fox Rd.
Laurel, MT 59044
406.628.2219
www.northernskies.com

Red Eagle Aviation
1880 Hwy. 93 S.
Kalispell, MT 59901
406.755.2376
www.redeagleaviation.com

Rocky Mountain College
1511 Poly Dr.
Billings, MT 59102
406.657.1000
aviation.rocky.edu

Summit Aviation LLC
490 Gallatin Field Rd.
Belgrade, MT 59714
406.388.8359
www.flysummit.net

NEBRASKA

University of Nebraska at Kearney
905 W. 25th St.
Kearney, NE 68849
308.865.8309
www.unk.edu/aviation

University of Nebraska – Omaha
6001 Dodge St.
Omaha, NE 68182
402.554.2800
ai.unomaha.edu

NEVADA

Monarch Sky LLC
1420 Jet Stream Dr., Ste. 105
Henderson, NV 89052
702.631.0386
www.monarchsky.com

NEW HAMPSHIRE

Daniel Webster College
20 University Dr.
Nashua, NH 03063
603.577.6570
www.dwc.edu

NEW JERSEY

Aero Prep Institute
60 Fostertown Rd.
Medford, NJ 08055
609.267.7673
aeroprep@flyingwairport.com
www.flyingwairport.com/flight_school.dws

Air Fleet Training Systems, Inc.
35 Wright Way
Fairfield, NJ 07004
973.575.8220
airfleettraining1@hotmail.com
www.airfleettraining.com

American Flyers
50 Airport Rd., Ste. 120
Morristown, NJ 07960
973.267.3223
sam@af.tv
www.americanflyers.net

Century Air, Inc.
19 Wright Way
Fairfield, NJ 07004
800.486.2368
info@centuryair.com
www.centuryair.com

Mercer County Community College

1200 Old Trenton Rd.
West Windsor, NJ 08550
609.586.4800
aviation@mccc.edu
www.mccc.edu/aviation

Ocean Aire

1514 Airport Rd.
Toms River, NJ 08754
732.797.1077
oceanaire@oceanaire.net
www.oceanaire.net

Privilege Aero, LLC

1034 Millstone River Rd.
Hillsborough, NJ 08844
732.803.2449
www.privilegeaero.com

Sky Training, LLC

126 Airport Rd.
West Milford, NJ 07480
973.728.1143
www.discovertheskies.com

NEW MEXICO

Bode Aviation, Inc.

7401 Paseo Del Volcan N.W.
Albuquerque, NM 87120
505.884.4530
info@flybode.com
www.flybode.com

NEW YORK

Academy of Aviation, LLC
7150 Republic Airport
Main Terminal Room 101
Farmingdale, NY 11735
631.777.7772
fly@academyofaviation.com
www.academyofaviation.com

Dowling College
150 Idle Hour Blvd.
Shirley, NY 11769
800.369.5464
www.dowling.edu

Dunkirk Aviation Flight School
3389 Middle Rd.
Dunkirk, NY 14048
716.366.6938
flightschool@dkk.com
www.dkk.com/flight3.html

East Hill Flying Club
62 Brown Rd.
Ithaca, NY 14850
607.257.1313
flyeasthill@gmail.com
www.ehfc.net

Farmingdale State College
Lupton Hall, Rte. 110
Farmingdale, NY 11735
631.420.2308
kamal.shahrabi@farmingdale.edu
www.farmingdale.edu

Heritage Flight Academy
2075 Smithtown Ave.
Ronkonkoma, NY 11779
800.786.9060
www.heritageflightacademy.com

Jamestown Community College
525 Falconer St.
Jamestown, NY 14702
716.338.1000
www.sunyjcc.edu/piloting

Prior Aviation Service, Inc.
50 N. Airport Dr.
Buffalo, NY 14225
716.633.1000
www.prioraviation.com

Rochester Air Center
1313 Scottsville Rd.
Rochester, NY 14624
585.328.8839
info@rochesterair.com
www.rochesterair.com

Schenectady County Community College
78 Washington Ave.
Schenectady, NY 12305
518.381.1200
www.sunysccc.edu

Vaughn College of Aeronautics
60-01 23rd Ave.
Flushing, NY 11369
718.429.6600
www.vaughn.edu

York College/CUNY Aviation Institute
94-20 Guy R. Brewer Blvd.
Jamaica, NY 11451
718.262.2829
www.york.cuny.edu/aviation

NORTH CAROLINA

Burlington Aviation
3441 N. Aviation Dr.
Burlington, NC 27215
336.227.1278
www.burlingtonaviation.com

Caldwell Community College & Technical Institute
2855 Hickory Blvd.
Hudson, NC 28638
828.726.2387
dsullivan@cccti.edu
www.cccti.edu

Dillon's Aviation
1105 N. Memorial Dr.
Greenville, NC 27834
252.757.1841
flightschool@dillonsaviation.com
www.dillonsaviation.com

Elizabeth City State University
1704 Weeksville Rd.
Elizabeth City, NC 27909
252.335.3443
jasharp@mail.ecsu.edu
www.ecsu.edu

Guilford Tech Community College
601 High Point Rd.
Jamestown, NC 27282
336.334.4822
amfloyd@gtcc.edu
www.gtcc.edu

Jacksonville Flying Service
278 Albert Ellis Airport Rd.
Richlands, NC 28574
910.324.2500
jfcassidy@gmail.com
www.flyjfs.com

Lenoir Community College
231 Hwy. 58 S.
Kinston, NC 28502
252.527.6223
jjennings@lenoircc.edu
www.lenoircc.edu

Tradewind Aviation International, LLC
820 Aviation Dr.
New Bern, NC 28562
252.636.0716
tradewind@coastalnet.com
www.tradewind-aviation.com

Wayne Community College
3000 Wayne Memorial Dr.
Goldsboro, NC 27534
919.739.6824
www.waynecc.edu/aviation

WNC Aviation LLC

21 Aviation Way
Fletcher, NC 28732
828.650.6540
athomas@wncaviation.com
www.wncaviation.com

NORTH DAKOTA

University of North Dakota

2806 Airport Dr.
Grand Forks, ND 58203
701.777.2791
flyund@aero.und.edu
www.aero.und.edu

OHIO

American Winds Flight Academy

1600 Triplett Blvd., Ste. C
Akron, OH 44306
330.733.2500
www.TeachMeToFly.com

Bowling Green State University College of Technology

1600 E. Wooster St.
Bowling Green, OH 43402
419.372.7563
jmcderm@bgsu.edu
www.bgsu.edu

Flamingo Air Academy

262 Wilmer Ave., Ste. 1
Cincinnati, OH 45226
513.321.7465
smcgee@flamingoairacademy.com
www.flamingoairacademy.com

Kent State University
119 Van Deusen Hall
Stow, OH 44224
330.672.2892
www.kent.edu

Ohio University
751 Columbia Rd.
Albany, OH 45701
740.597.2626
www.ohio.edu/aviation

Premier Flight Academy, ltd.
1535 Exeter Rd.
Akron, OH 44306
330.734.3660
www.premierflightacademy.com

Sporty's Academy, Inc.
2001 Sporty's Dr.
Batavia, OH 45103
513.735.9100
www.sportysacademy.com

The Ohio State University
164 W. 19 Ave.
Columbus, OH 43210
614.292.2405
www.aviation.osu.edu

University of Cincinnati Clermont College
4200 Clermont College Dr.
Batavia, OH 45103
513.735.5319
www.ucclermont.edu

Wright-Patterson AFB Aero Club
Bldg. 153, Area C
Wright-Patterson AFB, OH 45433
937.257.7714
joseph.callewaert@wpafb.af.mil
www.wpafbaeroclub.com

OKLAHOMA

Chickasha Wings, Inc.
2376 Beechcraft Rd.
Chickasha, OK 73018
405.244.8359
mitch@chickashawings.com
www.chickashawings.com

Oklahoma Aviation
7200 N.W. 63rd, Ste. 121
Bethany, OK 73008
405.787.4568
info@okav.co
www.oklahomaaviation.com

Oklahoma State University
339 Willard Hall
Stillwater, OK 74078
405.744.2739
terry.hunt@okstate.edu
www.flyosu.okstate.edu

Riverside Flight Center
241 Cessna Dr.
Tulsa, OK 74132
918.298.3164
info@riversideflightcenter.com
www.riversideflightcenter.com

Southeastern Oklahoma State University
10 Waldron Dr.
Durant, OK 74701
800.435.1327
dconway@se.edu
aviation.se.edu

Southern Nazarene University
6729 N.W. 39th Expy.
Bethany, OK 73008
405.789.6400
admissions@snu.edu
www.snu.edu

Spartan College of Aeronautics & Technology
8820 E. Pine St.
Tulsa, OK 74158
918.836.6886
www.spartan.edu

University of Oklahoma
1700 Lexington
Norman, OK 73069
405.325.7231
patriciamcgee@ou.edu
www.aviation.ou.edu

Western Oklahoma State College Pionair Flying Club
2801 N. Main
Altus, OK 73521
580.477.2000
chrystal.overton@wosc.edu
www.wosc.edu

OREGON

Aurora Aviation
22785 Airport Rd. N.E.
Aurora, OR 97002
503.678.1217
office@auroraaviation.com
www.AuroraAviation.com

Central Oregon Community College
2600 N.W. College Way
Bend, OR 97701
541.383.7700
aviation.cocc.edu

Corvallis Aero Service
5695 S.W. Airport Way
Corvallis, OR 97333
541.753.4466
john@corvallisaero.net
www.corvallisaeroservice.com

Gorge Winds Aviation, Inc.
920 N.W. Perimeter Way
Troutdale, OR 97060
503.665.2823
marv@gorgewindsinc.com
www.gorgewindsinc.com

Hillsboro Aviation, Inc.
911 N.W. Graham Rd.
Troutdale, OR 97060
503.489.1142
info@hillsboroaviation.com
www.hillsboroaviation.com

Lane Community College
28715 Airport Rd.
Eugene, OR 97402
541.463.4195
flight@lanecc.edu
www.lanecc.edu/flight

Portland Community College
17705 N.W. Springville Rd.
Portland, OR 97217
971.722.7256
erin.firecoat@pcc.edu
www.pcc.edu/fly

Professional Air
63132 Powell Butte Rd.
Bend, OR 97701
541.388.0019
info@proairservices.com
www.proairservices.com

Willamette Aviation Services, LLC
23115 Airport Rd. N.E.
Aurora, OR 97002
503.678.2252
info@WillametteAir.com
www.WillametteAir.com

PENNSYLVANIA

Ace Pilot Training, Inc.
600 Hayden Cr.
Allentown, PA 18109
610.264.1105
info@acepilot.com
www.acepilot.com

Aero-Tech Services, Inc.

500-P Airport Rd.
Lititz, PA 17543
717.394.2675
info@aerotechservices.com
www.aerotechservices.com

Aviation Certification & Education Solutions, Inc.

7 Piper St., Ste. 1
Beaver Falls, PA 15010
724.891.2237
fly@acesbvi.com
www.acesbvi.com

Community College of Allegheny County

1750 Clairton Rd. (PA Rte. 885)
West Mifflin, PA 15122
412.469.6222
glabuskes@ccac.edu
www.ccac.edu

Community College of Beaver County

125 Cessna Dr.
Beaver Falls, PA 15010
724.775.8561
carmen.romeo@ccbc.edu
www.ccbc.edu

Gateway Aviation, Inc.

1730 Vultee St.
Allentown, PA 18103
610.797.7942
info@gateway-aviation.com
www.gateway-aviation.com

Haski Aviation, Inc.
406 Frank Farone Dr.
New Castle, PA 16101
724.652.5546
haski@pathway.net
www.haski.com

Hortman Aviation Services Inc.
9800 Ashton Rd.
Philadelphia, PA 19114
215.969.0311
www.hortmanaviation.com

Lehigh Carbon Community College
4525 Education Park Dr.
Schnecksville, PA 18078
610.264.7089
www.lccc.edu

Marywood University
2300 Adams Ave.
Scranton, PA 18509
570.348.6211
www.marywood.edu

Moore Aviation, Inc.
7 Piper St.
Beaver Falls, PA 15010
724.843.4800
www.mooreaviationflighttraining.com

Moyer Aviation, Inc.
3800 Sullivan Trail
Easton, PA 18040
610.258.0473
www.moyeraviation.com

Pittsburgh Flight Training Center

Hangar 7
Lebanon Church Rd.
West Mifflin, PA 15122
412.466.1111
www.pghflight.com

Pittsburgh Institute of Aeronautics

5 Alleghany County Airport
West Mifflin, PA 15122
800.444.1440
pia.edu/aviation.html

Tech Aviation Flight School

101 Hangar Rd.
Avoca, PA 18641
570.414.0223
www.techaviationflightschool.com

RHODE ISLAND

Cindtronix Aviation

210 Airport St.
North Kingstown, RI 02852
401.667.7580
www.cindtronix.com

Horizon Aviation, Inc.

660 Airport Rd.
Warwick, RI 02886
401.736.5115
www.horizonaviation.com

SOUTH CAROLINA

ASG Flight Training Academy
552 Perimeter Rd.
Greenville, SC 29605
866.373.2853
info@aserve.com
www.avserve.com

SOUTH DAKOTA

Aberdeen Flying Service
4430 U.S. 12
Aberdeen, SD 57401
605.225.1384
afsbc@nvc.net
www.aberdeenflyingservice.com

Quest Aviation, Inc.
4440 E. Hwy. 12
Aberdeen, SD 57401
877.225.8058
questkb@nrctv.com
www.questaviationinc.com

South Dakota State University
201 W. 2nd St. S.
Brookings, SD 57007
605.688.4121
www.sdstate.edu/cs/programs/aviation.cfm

TENNESSEE

Downtown Aviation
2787 N. Second St.
Memphis, TN 38103
901.353.9151
www.downtownaviation.com

Middle Tennessee State University
1301 E. Main St.
Murfreesboro, TN 37132
615.898.2788
asymonds@mtsu.edu
aerospace.web.mtsu.edu/home.htm

Tennessee State University
3500 John A. Merritt Blvd.
Nashville, TN 37209
615.963.5378
www.tnstate.edu/ait

Wings of Eagles
276 Doug Warpoole Rd.
Smyrna, TN 37167
615.355.0033
www.wingsmqy.com

TEXAS

Air Safety Flight Academy
5318 Airport Dr.
Denison, TX 75020
www.airsafetyacademy.com

Alpha Tango Flying Services, Inc.
1331 Northern Blvd.
San Antonio, TX 78216
210.828.4480
www.alphatango-flying.com

Austin Academy of Aviation
4309 General Aviation Ave.
Austin, TX 78719
512.385.2880
www.flyaus.com

Aviation Academy of America, Inc.
1764 Entrance Dr.
New Braunfels, TX 78130
830.629.1700
info@academytx.com
www.flyskyhawk.com

Aviator Air Centers, Inc.
3122 Great S.W. Pkwy.
Grand Prairie, TX 75052
972.988.8609
sales@aviatorair.com
www.aviatorair.com

Baylor University
1311 S. 5th St.
Waco, TX 76706
254.710.3563
timothy_compton@baylor.edu
www.baylor.edu/aviation

Brazos Valley Flight Services
1680 George Bush Dr. W., Ste. 2
College Station, TX 77845
979.260.6322
www.bvfs.net

Central Texas College
6200 W. Central Texas Expy.
Killeen, TX 76540
254.526.1241
ctc.aviation@ctcd.edu
www.ctcd.edu/aviation

Decatur Jet Center
351 Airport Rd.
Decatur, TX 76234
940.627.1050
www.decaturjetcenter.com

Delta Qualiflight Aviation
151 Meacham Airport Cr., Bldg. 4 N.
Fort Worth, TX 76106
817.626.6300
www.deltaqualiflight.com

Flying Tigers
11800 Kirk Blvd.
Houston, TX 77034
713.649.5400
www.goflyingtigers.com

Gold Star Aviation
200 Commander Rd. H2NC
Fort Worth, TX 76106
817.624.7320
www.goldstarav.com

Texas State Aviation
2049 Airport Dr.
San Marcos, TX 78666
512.396.2212
www.tsaviation.com

Hi-Tech Aviation, Inc.
9306 Max Conrad
Spring, TX 77379
281.370.5235
www.hitechaviation.com

Hub City Aviation, Inc.
6004 N. Cedar Ave., Ste. A
Lubbock, TX 79403
806.687.1070
info@hubcityaviation.com
www.hubcityaviation.com

LeTourneau University
2100 S. Mobberly Ave.
Longview, TX 75602
800.759.8811
www.letu.edu

Marcair, Inc.
11310 Cleveland Gibbs Rd.
Roanoke, TX 76262
817.430.0005
learntofly@marcairaviation.com
www.marcairaviation.com

Monarch Air
4580 Claire Chennault
Addison, TX 75001
972.931.0345
www.monarchair.com

Mountain View College
4849 W. Illinois Ave.
Dallas, TX 75211
214.860.8763
aviation.mvc.dcccd.edu

Palo Alto College
1400 W. Villaret Blvd.
San Antonio, TX 78224
210.486.3077
www.alamo.edu/pac

Pilots Choice Aviation
209 Corsair Dr.
Georgetown, TX 78628
512.869.1759
www.pilotschoice.com

Pro Aircraft Flight Training
137 Aviator Dr.
Fort Worth, TX 76179
817.439.3033
www.pro-aircraft.com

Pro Flite Aviation Services
8411 Mission Rd.
San Antonio, TX 78214
210.922.1228
www.profliteaviation.com

San Jacinto College
8060 Spencer Hwy.
Pasadena, TX 77505
281.476.1501
www.sanjac.edu

Sky Safety, Inc.
8603 Mission Rd.
San Antonio, TX 78214
210.921.2504
www.skysafety.com

Skymates Inc.
5080 S. Collins St., Ste. 104
Arlington, TX 76018
817.472.8307
www.skymates.com

Slipstream Aviation
5435 Saturn Dr.
Dallas, TX 75237
817.658.5988
mischa@flyslipstream.com
www.flyslipstream.com

Streamline Aviation
4321 General Aviation Ave.
Austin, TX 78719
800.575.5098
www.streamlineair.com

Texas A & M University – Central Texas
1901 S. Clear Creek Rd.
Killeen, TX 76549
254.519.5469
www.tarleton.edu/centraltexas/departments/aviation

Texas Southern University
3100 Cleburne Dr.
Houston, TX 77004
713.313.1395
www.tsu.edu

TSTC Harlingen
1902 N. Loop 499
Harlingen, TX 78550
956.364.4756
www.harlingen.tstc.edu

TSTC Waco
3801 Campus Dr.
Waco, TX 76705
800.792.8784
www.waco.tstc.edu

TSTC West Texas
2850 Airport Blvd.
Abilene, TX 79602
325.671.8009
www.tstcairacademy.com

United Flight Systems, Inc.
20119A Stuebner Airline
Spring, TX 77379
281.376.0357
www.unitedflight.com

U.S. Flight Academy International Inc.
2000 Rickabaugh Dr., Bldg. 1106
Big Spring, TX 79720
432.264.7335
www.usflightacademy.org

UTAH

Salt Lake Community College
4600 S. Redwood Rd.
Salt Lake City, UT 84123
801.957.4448
www.slcc.edu

Utah State University
6000 Old Main Hill
Logan, UT 84322
435.753.4289
www.usu.edu/aviation

Utah Valley University
1185 Mike Jense Pkwy.
Provo, UT 84606
888.901.7192
www.flyuvu.com

Westminster College
1840 S. 1300 E.
Salt Lake City, UT 84105
800.748.4753
www.westminstercollege.edu

VERMONT

Lakeview Aviation
2628 Airport Rd.
Newport, VT 05855
802.334.5001
www.lakeviewaviation.com

North Ramp Aviation
1150 Airport Dr., Ste. 101
South Burlington, VT 05403
802.865.4366
www.northrampaviation.com

Sugarbush Soaring Association
2355 Airport Rd.
Warren, VT 05674
802.496.2290
soar@sugarbushsoaring.com
www.sugarbushsoaring.com

VIRGINIA

Averett University
420 W. Main St.
Danville, VA 24541
434.791.5600
www.averett.edu

Aviation Adventures

1001 Sycolin Rd. S.E.
Leesburg, VA 20175
703.777.6030
kjyo@aviationadventures.com
www.aviationadventures.com

Dulles Aviation Inc.

10501 Observation Rd.
Manassas, VA 20110
703.361.2171
Info@DullesAviation.com
www.DullesAviation.com

Falwell Aviation Jet Center

310 Hangar Rd.
Lynchburg, VA 24502
434.237.7002
soliver@flyfreedomaviation.com
www.flyfalwell.com

Hampton University

100 E. Queen St.
Hampton, VA 23668
757.727.5000
www.hamptonu.edu

Horizon Flight Center

2801 Airport Dr.
Chesapeake, VA 23323
757.421.9000
www.horizonflightcenter.com

JLS Aviation, LLC
3380 Shannon Airport Cr.
Fredericksburg, VA 22408
540.374.1715
jlsaviation@cox.net
www.jlsaviation.info

Langley Air Force Base Aero Club
308 Emmons St.
Langley AFB, VA 23665
757.766.1347
www.langleyaeroclub.org

Liberty University
1971 University Blvd.
Lynchburg, VA 24502
434.582.2183
aviation@liberty.edu
www.liberty.edu/aviation

Pilot Makers, LLC
902-D Bland Blvd.
Newport News, VA 23602
757.232.0278
james@flyhampton.com
www.flyhampton.com

Virginia Aviation
970 Airport Rd.
Lynchburg, VA 24502
434.237.8420
customerservice@virginiaaviation.com
www.virginiaaviation.com

WASHINGTON

Avian Flight Center
8900 State Hwy. 3 S.W.
Port Orchard, WA 98367
866.263.2359
www.avianflight.com

Big Bend Community College
7662 Chanute St.
Moses Lake, WA 98837
509.793.2241
aviation@bigbend.edu
www.bigbend.edu/aviation

Central Washington University
400 E. University Way
Ellensburg, WA 98926
509.963.2364
aviation@cwu.edu
www.cwu.edu/~flight

Clover Park Technical College
4500 Steilacoom Blvd. S.W.
Lakewood, WA 98499
253.589.5800
bill.coyner@cptc.edu
www.cptc.edu

Galvin Flying Services
7001 Perimeter Rd.
Seattle, WA 98108
800.351.5401
www.galvinflying.com

Green River Community College
12401 S.E. 320th St.
Auburn, WA 98092
253.833.9111
www.greenriver.edu

Inter-State Aviation, Inc.
4800 Airport Complex N.
Pullman, WA 99163
509.332.6596
douggadua@inter-stateaviation.com
www.inter-stateaviation.com

Jetstream Aviation, Inc.
45 Airport Way
Dallesport, WA 98617
509.767.0014
www.jetstreamaviation.com

Midstate Aviation, Inc.
1101 Bowers Rd.
Ellensburg, WA 98926
509.962.7850
midstateaviation@fairpoint.net
www.midstateaviation.net

Moody Aviation
6719 E. Rutter Ave.
Spokane, WA 99212
509.535.4051
moody.aviation@moody.edu
www.moodyav.org

Northway Aviation
10108 32nd Ave. W., Bldg. C-3
Everett, WA 98204
425.742.7003
contact@northwayaviation.com
www.northwayaviation.com

Northwest Aviation College
506 23rd St. N.E.
Auburn, WA 98002
253.854.4960
info@NACstaff.com
www.NorthwestAviationCollege.edu

Pavco Flight Center
1110 26th Ave. N.W.
Gig Harbor, WA 98335
253.851.5577
school@flypavco.com
www.flypavco.com

Regal Air
10217 31st Ave. W., Bldg. C51
Everett, WA 98204
425.353.9123
fly@regalair.com
www.regalair.com

Snohomish Flying Service
9900 Airport Way
Snohomish, WA 98296
360.568.1541
snofly@harveyfield.com
www.snohomishflying.com

Spokane Falls Community College
3410 W. Fort George Wright Dr.
Spokane, WA 99224
509.533.3350
jcanty@aero.und.edu
www.spokanefalls.edu/academic/aviation

Walla Walla College
204 S. College Ave.
Walla Walla, WA 99324
800.541.8900
loury.duffy@wallawalla.edu
www.wallawalla.edu

Wings Aloft
8467 Perimeter Rd. S.
Seattle, WA 98108
206.763.2113
info@wingsaloft.com
www.wingsaloft.com

WEST VIRGINIA

Elkins Aviation
1000 Airport Rd.
Elkins, WV 26241
304.460.1192
elkins_aviation@live.com
www.elkinsaviation.com

Fairmont State University
1201 Locust Ave.
Bridgeport, WV 26554
304.842.8300
thomas.stose@fairmontstate.edu
www.fairmontstate.edu

Mountain State University
609 S. Kanawha St.
Beckley, WV 25801
866.367.6781
www.aviation.mountainstate.edu

WISCONSIN

Fox Valley Technical College
150 N. Campbell Rd.
Oshkosh, WI 54902
920.735.5600
huss@fvtc.edu
www.fvtc.edu

Osceola Aerosport, LLC
495 S. Cascade St.
Osceola, WI 54020
715.294.4500
fly@osceolaaero.com
www.osceolaaero.com

Wisconsin Aviation, Inc.
N6491 State Hwy. 26
Juneau, WI 53039
920.386.2402
wisav@wisconsinaviation.com
www.wisconsinaviation.com

WYOMING

Crosswind Aviation, Inc.
7548 Fuller St.
Casper, WY 82604
307.472.4359
info@crosswindavi.com
www.crosswindavi.com

Wings of Wyoming
3803 Evans Ave.
Cheyenne, WY 82001
866.786.2359
info@wingsofwyoming.com
www.wingsofwyoming.com

Online Resources

Many aviation associations, websites, and periodicals offer information on flight schools and aviation scholarships, including the Faa.gov website. The following are some of the more well-known, but many other sources exist. Remember, thorough research is essential when choosing schools and programs and for learning more about the industry.

Aviation Associations and Pilot Unions

Air Line Pilots Association, International
www.alpa.org

Air Transport Association of America, Inc.
www.airlines.org

Aircraft Owners and Pilots Association
www.aopa.org

Allied Brotherhood of Teamsters Union Airline Pilots
www.ibt1224.org

Allied Pilots Association
www.alliedpilots.org

AirTran Pilots Association
airtranpilots.org

Alpha Eta Rho – International Aviation Fraternity
www.alphaetarho.org

Airplane Owners and Pilots Association
www.aopa.com

Coalition of Airline Pilots Associations
www.capapilots.org

Experimental Aircraft Association
www.eaa.org

FedEx Pilot Association
fdxpilot.com

Independent Pilots Association (UPS)
www.ipapilot.org

International Federation of Air Line Pilots' Associations
www.ifalpa.org

National Business Aviation Association
www.nbaa.org

Regional Airline Association
www.raa.org

Southwest Airlines Pilots' Association
www.swapa.org

University Aviation Association
www.uaa.aero

Women in Aviation, International
www.wai.org

Industry Publications

Aero.com – an online alphabetical listing of worldwide aviation magazines
www.aero.com

Air & Space Magazine
www.airspacemag.com

***AOPA Flight Training Magazine* for student pilots**
flighttraining.aopa.org/magazine

***AOPA Pilot, AOPA ePilot*, AOPA's Airport Directory**
www.aopa.org/publications

***Aviation Week* – Aviation, defense, and space news**
www.aviationweek.com

***Professional Pilot* – for corporate pilots**
www.propilotmag.com

Thirty Thousand Feet – an online listing of aviation-related magazines and webzines
www.thirtythousandfeet.com/magazine.htm

Other Aviation-Related Websites

Advisory Circulars – a resource for locating Advisory Circulars worldwide
advisorycirculars.com

Airline Pilot Central – information and articles for airline pilots

www.airlinepilotcentral.com

AirNav.com – various types of aviation information

www.airnav.com

AOPA Flight Training – online resources for student pilots

flighttraining.aopa.org

The Aviation Directory – "If you can't find it in avhome.com, it probably doesn't exist!"

www.Avhome.com

Aviation Jobs Online TM

www.aviationjobsonline.com

The 'Aviator' Web Site – information and merchandise for pilots

www.aviatorwebsite.com

AvScholars – information on aviation colleges, flight training, scholarships, etc.

www.avscholars.com

TheCFI.com – information for certified flight instructors, plus links to other sites

www.thecfi.com

Climb to 350 – aviation career connections

www.climbto350.com

Federal Aviation Administration – information about almost every area of aviation, federal rules and advisories, flight schools and scholarships, etc.
www.faa.gov

FlightSchoolList.com – lists of flight schools by state, news, forums, articles, games
www.flightschoollist.com

The Flying Trader – aviation classified ads
www.theflyingtrader.com

Landings – aviation news, articles, directories
www.landings.com

Pilotfriend – pilot resources and aviation weather for general aviation
www.pilotfriend.com

PilotJourney – information on learning to fly
www.pilotjourney.com

Thirty Thousand Feet – an aviation directory with links to aviation web pages, aviation news, and other sources of commercial, military, and general aviation information
www.thirtythousandfeet.com

Index

A

B

C

D-E

F

G-I

J-K

L

M

N-O

P

R

S

T

V-W-X-Y

About the Author

Jodi Wheatley

Born and raised in Texas, Jodi Wheatley earned her bachelor's degree in TV and Film Production from the Journalism, Broadcasting, and Film school at Trinity University in San Antonio. After many years working in local television – writing, producing and directing commercials, public service announcements, station promotions, and marketing videos – she expanded into technical/communications writing for Texas State Technical College West Texas. Writing and editing online course content (including fiber optics, hydraulics, cutting and welding, electricity, applied math and other subjects) allowed her to do the two things she enjoys most: learning and helping others learn.

Established in 2004, TSTC Publishing is a provider of high-end technical instructional materials and related information to institutions of higher education and private industry. "High end" refers simultaneously to the information delivered, the various delivery formats of that information, and the marketing of materials produced. More information about the products and services offered by TSTC Publishing may be found at its website: publishing.tstc.edu.

TechCareers Series

TSTC Publishing launched the TechCareers series with *Biomedical Equipment Technicians* in 2008. TSTC Emerging Technologies initially underwrote the series, created to inform the public about existing technologies and those to come.

Emerging Technologies also provided funding for 500 copies of each book in the series to be distributed throughout Texas to high school career and technical education counselors. In addition to *Biomedical Equipment Technicians*, the series includes *Automotive Technicians*, *Wind Energy*, *Avionics* and *Computer Gaming Programmers & Artists*. Forthcoming titles include: *Aviation Maintenance*, *Aviation Pilots* and *Graphic Design*. For information about Emerging Technologies, go to forecasting.tstc.edu.

Every TechCareers book features:

- Detailed overviews of career pathways, skill sets and educational requirements
- Profiles of professionals, experts, employers, current students and instructors
- Program listings, sample degree plans and additional industry resources
- Salary ranges and benefits

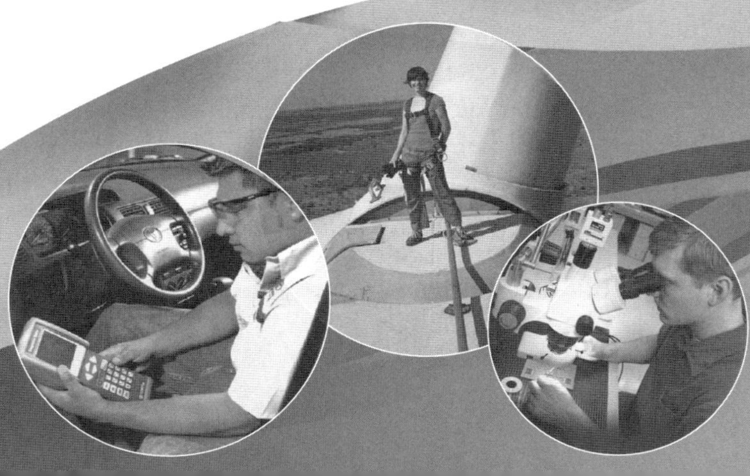

TechCareers: Wind Energy

By Mike Jones

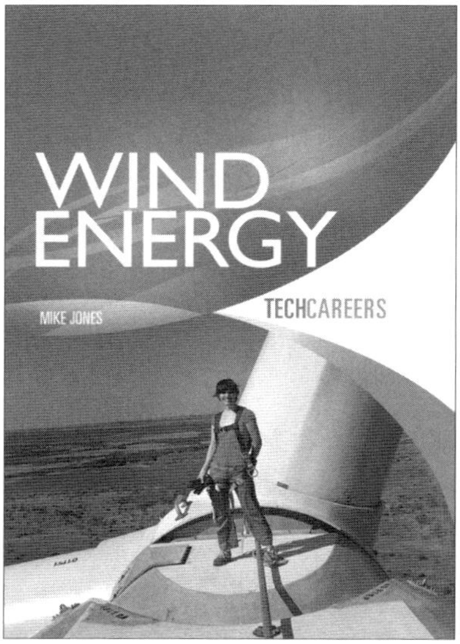

$14.95 Softback
ISBN 978-1-934302-55-2
$9.99 ebook
ISBN 978-1-936603-02-2
1st edition
August 2010

Kindle edition available at Amazon.com

> "With increasing demand for wind power, there is already a pressing need for wind technicians to maintain the wind turbines."
>
> ***-TechCareers: Wind Energy***

TechCareers: Wind Energy describes the current wind energy market and the explosive growth of the energy field. Due to current interest in green energy, wind energy is now becoming a popular energy source in countries across the world. Increased demand for expert wind energy professionals is expected to continue. This book describes the jobs needed to support this growing career field and the education and necessary skills for success.

TechCareers: Avionics

By Helen Ginger

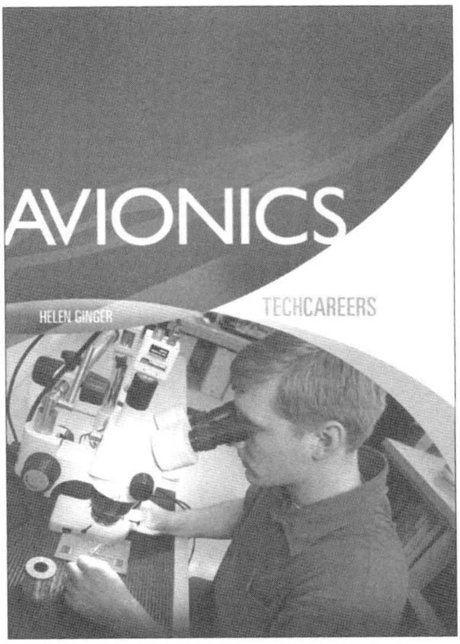

$14.95 Softback
ISBN 978-1-934302-47-7
1st edition
October 2009

> "If you're just beginning to think about this field and a career in avionics, now is the opportune time to begin your training ... the good news for those now considering this field is that the super techs are reaching retirement age, creating a job gap and a wide open door for new techs to step through."

-*TechCareers: Avionics*

TechCareers: Avionics gives a clear description of the growing need for avionics technicians. Avionics deals with the maintenance and repair of all flight instruments, including flight control, weather radar, and missile control. With many of the current avionics technicians reaching retirement, airlines will be looking for new techs to take their places. Featuring necessary information about job opportunities, training, and educational requirements, *TechCareers: Avionics* is a valuable tool for prospective students entering the field.